MATHS FOI S

BY

D. GALLOWAY M.A.
Biology Department

A. M. GENT M.A.
Head of Mathematics Department

J. HUTTON Ph. D.
Chemistry Department

V. PAYNE Ph. D.
Head of Physics Department

MALVERN GIRLS' COLLEGE

MILLS & BOON LIMITED, LONDON

First published 1980 in Great Britain
by Mills & Boon Limited,
17–19 Foley Street
London, W1A 1DR

ISBN 0 263 06368 2

Filmset and printed in Great Britain by
Thomson Litho Ltd, East Kilbride, Scotland
and bound by Hunter & Foulis Ltd, Edinburgh

PREFACE

Students of Biology, Chemistry, Physics, need some mathematical understanding if these subjects are to be mastered, and yet many do not study Mathematics as a separate course to a similar level. Even those who do can find difficulty in relating their maths to practical needs in the laboratory or to theory. This text is the outcome of close collaboration between four teachers (of Biology, Mathematics, Chemistry and Physics) who have written a text—brief, but sound—to supply the mathematical needs of such students.

The course was developed initially in teaching sixth forms working for the G.C.E. Advanced level examinations, but the text is not tied to such syllabuses and should be equally useful wherever, in school, college or university, greater mathematical insight is needed by science students at approximately this stage in their development. The topics discussed cover as wide a range as possible, and worked examples demonstrate the relevance of the work to the sciences as the student goes along. Included are revision of some elementary processes; practice of techniques and their applications; new material germane to one or other of the sciences; broadening of the Mathematical vocabulary to enable students trained in traditional ways to understand the modern idiom and vice versa; critical appraisal of scientific results. The background knowledge assumed is approximately that needed for G.C.E. O level in Mathematics, and experience has shown the course is helpful to students over a wide ability range.

There are helpful additions to the syllabuses of the Examining Boards, giving details of the Mathematics expected for Biology, Chemistry and Physics.

What has been included can be taught in one year, assuming that approximately one-and-a-half hours are available each week. The authors are very conscious that this deliberate restriction in content has meant that scant attention is paid to some matters of importance, and that others, also important, have been left out. The order of the chapters is that which has proved most suitable for the sixth form science students who have been using the course.

ACKNOWLEDGEMENTS

We would like to acknowledge our debt to Malvern Girls' College for the active support which, to give only one instance, allowed four or five teachers to participate simultaneously in the course at its inception, and for the interest, comments and questions from pupils which were always stimulating. John Galloway, Hubert Gent and John Payne have all contributed beyond the line of duty. Thanks also to Avril Stevens for her careful typing.

The Joint Matriculation Board has been helpful in allowing the use of Advanced level questions.

We are grateful to the Literary Executor of the late Sir Ronald A. Fisher, F.R.S., to Dr. Frank Yates, F.R.S. and to Longman Group Ltd., London, for permission to reprint Tables IV and VII from their book *Statistical Tables for Biological, Agricultural and Medical Research.* (6th edition, 1974.)

Finally, we should like to express our thanks to Professor D. J. Waddington of the University of York for his encouragement and many helpful comments.

D. Galloway
A. M. Gent
J. Hutton
V. Payne

Malvern Girls' College

1

AIDS TO CALCULATION

In the laboratory, tables of logarithms and slide rules have long been used, and now many students have electronic calculators. All these have their place, but before discussing their use some general principles must be considered.

ACCURACY

The design of an experiment and the instruments used in it dictate the accuracy of the results.

Example 1

A tape measure is used to measure the radius of a circle as 2 m. We record this as 2.00 m indicating that the measurement can be relied on to the nearest centimetre. The circumference is needed; $C = 2\pi r$ and a calculator with a π button gives the answer as 12.56636. How do you record this? You might quote $C = 12.57$ m, to the nearest centimetre as before, but for reasons which will be discussed in chapter 2 we should write $C = 12.6$ m.

Example 2

Laser beams enable lengths to be measured extremely accurately over long distances, so that in an experiment using one to measure earth movements, 10532.094 m is a realistic result. Any calculations which ensued would need a computer or calculator working to 9 or more figures; logarithms or slide rule would be useless.

These two examples illustrate the need for care, not only in avoiding mistakes but in expressing results in a form appropriate to the degree of accuracy expected from the experiment. In this chapter much of the work is arithmetical and may not appear to be directly concerned with your Science course, but the techniques and definitions are very necessary for it. It is no good doing a quantitative experiment if you cannot handle the figures. Notice that the accuracy attainable in the calculating aid you use must always be considered; if it is too low you are not making full use of the experimental data you are processing.

SIGNIFICANT FIGURES AND DECIMAL PLACES

The first significant figure in a number is the first digit from the left other than 0; the number of decimal places in it is the number of digits after the decimal place.

Example 3

12.56636 has 7 significant figures and is quoted to 5 decimal places.

	No. of sig. fig.	*No. of dec. pl.*
12.56636	7	5
12.57	4	2
10532.094	8	3
0.000461	3	6

Correcting to a given number of decimal places is most easily explained by an example; suppose we wish to quote 0.000461 to 4 decimal places. Look at the 5th place, here 6. Ask "is $6 \geqslant 5$?" The answer is "yes", so we say $0.000461 \approx 0.0005$ to 4 dec. pl. ($\approx$ is used to mean approximately equal to). Similarly, to give 10532.094 to 3 significant figures, look at the 4th significant figure which is 3. Now $3 < 5$, so we say $10532.094 \approx 10500$ to 3 sig. fig. In words, this could be expressed as giving the answer to the nearest 100.

Example 4

Express each of the following numbers in the way indicated.

57.49 to 1 sig. fig;	60
0.0109 to 2 sig. fig;	0.011
3.075 to 2 decimal places;	3.08
0.0232 to 3 decimal places;	0.023.

STANDARD FORM OR SCIENTIFIC NOTATION

You might quote a wavelength of light as 4026×10^{-10} m. In standard form this is written 4.026×10^{-7} m.

Every number can be expressed in this way, as $x = n \times 10^p$, where $1 \leqslant n < 10$, and p is any integer (a whole number which is positive, negative or 0). This form is much used; numbers are quickly compared, the degree of accuracy is at once apparent, and errors are reduced because for instance, 700000000 is more likely to be copied incorrectly than 7×10^8. Where the facility called Scientific Notation is available on an electronic calculator, numbers are shown on display in standard form.

Example 5

Express 0.0000003105 and 7810000 in standard form.

$0.0000003105 = 3.105 \times 10^{-7}$,

$7810000 = 7.81 \times 10^6$.

ESTIMATES

You make estimates without thinking about it in everyday matters; "50 miles on the motorway—do it easily in under the hour" or "Concert £1, buses 30p—I should be all right if I take £2". In more complex calculations a routine is needed and this is based on standard form. An example will show the process clearly.

Example 6
Calculate K, where

$$K = \frac{0.0341 \times 2.23 \times 10^7}{1.075 \times 8692},$$

first making an estimate of the answer to 1 significant figure.

Step 1. Put all numbers into standard form;

$$K = \frac{3.41 \times 10^{-2} \times 2.23 \times 10^7}{1.075 \times 8.692 \times 10^3}.$$

Step 2. Simplify powers of 10 and correct all numbers to 1 significant figure;

$$K \approx \frac{3 \times 2 \times 10^2}{1 \times 9}$$

Step 3. Calculate K in this form to make your estimate;
$K \approx 70$.

Step 4. Use your chosen aid to calculate K; $K = 81.3825$.
$K = 81.4$ (using a slide rule).

Note: In the original expression for K two of the numbers are quoted to 3 sig. fig. This sets the limit on the accuracy attainable, which is therefore only to 3 sig. fig. and for which a slide rule is adequate.

LOGARITHMS

With an electronic calculator at hand you are unlikely to use logarithms as an aid to computation; you still need them though (there are some applications in chapter 4) and the theory is dealt with in chapter 3. We include here a summary of methods.

FINDING LOGARITHMS

With a calculator, use the key marked log (but check in the instruction booklet, or try with a number whose log you know). With tables, first put the number in standard form because tables list the logarithms of numbers from 1.000 to 9.999 only.

MULTIPLICATION

Add the logarithms $\log(xy) = \log x + \log y$.

DIVISION

Subtract the logarithms $\log(x/y) = \log x - \log y$.

POWERS AND ROOTS

Multiply the logarithm by the power; $\log(x^n) = n\log x$. Note that n may be positive or negative, an integer or indeed any real number, so that the same rule holds for raising to any power and for taking any root.

NEGATIVE LOGARITHMS, BAR NOTATION

The logarithm of a number between 0 and 1 is negative and a calculator will give log 0.2 as -0.69897. This is what you want for many purposes, but you may also express log 0.2 in the form $\bar{1}.30103$, meaning $-1+0.69897$, read as "bar one point..."

Exercise A

In numbers 1–17, first make an estimate of the answer (1 or 2 significant figures) then perform the calculation. Record the result to three significant figures.

1. 2.471×39.70
2. 496×2.52
3. 0.6354×0.0493
4. 7032×21.60
5. $29.63 \div 9.760$
6. $104300 \div 5761$
7. $8973 \div 8734$
8. $\dfrac{26.42 \times 0.8537}{9.027}$
9. $\dfrac{37.28}{62.42}$
10. $(3.086)^4$
11. $\sqrt[3]{93.64}$
12. $614 \div (73)^2$
13. $\dfrac{94.3}{16.0}$
14. $2\pi\sqrt{\dfrac{0.34}{9.80}}$
15. $(0.19)^{13}$
16. $4.73 \div 98.22$
17. $\sqrt[3]{0.1}$
18. Write down the logarithms (*not* using the bar notation) of π, 3.89, 0.196, 0.00472, 0.0381, $\frac{1}{4}$, $\frac{0.1}{60}$, 0.0001, 0.9, $\frac{1}{2.4}$.

SLIDE RULE

You may find your slide rule a very convenient aid; if so you will find that most of the examples in Exercise A can be done with it. Remember to estimate as you go along, and to check that you are achieving sufficient accuracy for your purpose.

ELECTRONIC CALCULATORS

The models available change almost from day to day, so no detailed instruction is possible here. If you are buying a new machine find out whether there are any restrictions imposed in the examinations for which you are preparing (programmables excluded, for instance). Many calculators now are so powerful that you need a good deal of practice to use them to the full although all of them are useful from the moment you switch on.

Batteries run out (even on liquid crystal display models), calculators need repair, but the experiment goes on, so do not allow yourself to lose all your skill in arithmetic.

Exercise B

The problems in this exercise are drawn from Physics, Chemistry and Biology and some require background knowledge for their solution. In the calculations apply the techniques discussed in this chapter.

1. Use the Bragg equation, $n\lambda = 2d \sin\theta$, to calculate the wavelength of the X-rays used in a simple crystal structure determination, given that a second order reflection ($n = 2$) occurred for $\theta = 14°27'$, and the interplanar spacing d is 284 pm.
2. Calculate the new volume that the following gases would occupy under s.t.p. conditions, 273 K and 760 mm Hg pressure, assuming ideal behaviour:

   (a) 6.8 dm^3 measured at 294 K and 670 mm Hg,
   (b) 10.3 dm^3 measured at 347 K and 750 mm Hg, and
   (c) 30 cm^3 measured at 260 K and 507 mm Hg.

3. Given that the K_b value for the dissociation of X in aqueous solution is 3.2×10^{-6} mol dm^{-3}, calculate the K_a value if $K_a \times K_b = 10^{-14}$ $mol^2\,dm^{-6}$.
4. A solution was made by mixing 10.4 g of an organic compound $A(M_r = 104)$ the solute, and 29 g of solvent $B(M_r = 58)$. Calculate the mole fraction of A and B, x_A and x_B, in the solution. If the vapour pressure of pure B is 120 mm Hg what is the vapour pressure P_B of B above the solution at the same temperature?

   $$\left(\frac{P_B^0 - P_B}{P_B^0} = x_A \text{ assuming ideal solution behaviour.}\right)$$

5. The rates of diffusion of gases under given conditions are inversely proportional to the square roots of their densities. If the rate of diffusion of a gas is a quarter that of another gas of relative molecular mass 14 under comparable conditions, then what is the relative molecular mass of the first gas?
6. Given that $F = 9.650 \times 10^4$ coulombs mol^{-1}, what is the charge carried by 1 mole of α-particles? (*N.B.* α-particles are helium nuclei having a mass of four, and a positive charge of two units i.e. ${}_2^4He^{2+}$.)
7. If the concentration of hydroxide ions is increased by a factor of 1000, what is the resulting change in the pH?
8. For a given mass of air, the original volume, V_1 is 120 cm^3 at a pressure, $P_1 = 74$ cm of mercury and temperature, $T_1 = 300$ K. Find the new volume, V_2, at the pressure, $P_2 = 76$ cm Hg and temperature $T_2 = 250$ K, using the equation $P_2 = \dfrac{P_1 V_1 T_2}{V_2 T_1}$.
9. A mass of air, occupying initially a volume, $V_1 = 222$ cm^3 at a pressure of $P_1 = 76.0$ cm Hg is expanded adiabatically to a volume $V_2 = 250$ cm^3. Calculate the new pressure, P_2, given that the equation for an adiabatic change is $P_2 = \dfrac{P_1 V_1^\gamma}{V_2^\gamma}$ where $\gamma = 1.40$ for air.
10. The temperature, $T_1 = 273$ K of a given mass of gas, occupying volume $V_1 = 97$ cm^3 is increased to $T_2 = 373$ K under adiabatic conditions.

Given that $V_2^{\gamma-1} = \dfrac{T_1 V_1^{\gamma-1}}{T_2}$ where $\gamma = 1.40$, find the new volume, V_2.

11. Calculate the gravitational force, F, on an object of mass, $m = 1.0\,\text{kg}$ at the surface of the earth, mean radius $6.4 \times 10^6\,\text{m}$ and mass, $M = 6.0 \times 10^{24}\,\text{kg}$, if the gravitational force, F, between the earth mass M, and an object of mass, m, is given by

$$F = \frac{GMm}{R^2}$$

where R is the distance of the object from the centre of the earth and the gravitational constant $G = 6.7 \times 10^{-11}\,\text{Nm}^2\,\text{kg}^{-2}$.

12. Calculate the period, T of a simple pendulum of length $l = 1.00\,\text{m}$ given that

$$T = 2\pi\sqrt{\frac{l}{g}} \quad \text{and} \quad g = 9.81\,\text{m}\,\text{s}^{-2}.$$

13. Einstein's photo electric equation is written as

$$h\nu - W_0 = E$$

where h is Planck's constant $= 6.6 \times 10^{-34}\,\text{J}$. Calculate the energy of emission, E, for the photo electron emitted when light of frequency, $\nu = 6.7 \times 10^{14}\,\text{Hz}$ is incident on the surface whose work function, $W_0 = 2.9 \times 10^{-19}$ joules.

Also, calculate the speed of emission, v, where v is given by the equation

$$v = \sqrt{\frac{2E}{m}}$$

where $m = 9.1 \times 10^{-31}\,\text{kg}$, the mass of the electron.

14. I.Q. (intelligence quotient) $= \dfrac{\text{mental age}}{\text{real age}} \times 100$

Calculate the I.Q. of the following people:

	Mental age		*Real age*	
	Years	*Months*	*Years*	*Months*
a	9	6	10	0
b	15	2	14	1
c	13	11	14	7
d	7	8	8	2
e	8	10	8	6

15. Assume that blood makes up 7% of the body's weight. If each of the following people donate one pint of blood, what percentage of their blood will each have given?

	Body weight
a	8 stone
b	9 stone
c	10 stone
d	11 stone
e	12 stone

N.B.
1 pint blood weighs 600 g.
1 stone = 6.35 kg.

16. 100 g Cheddar cheese contains 810 mg calcium.
100 g milk contains 120 mg calcium.
How much (a) cheese, (b) milk, would be needed to provide the whole daily requirement of 500 mg calcium?
17. A fairly active man requires 12,600 kJ energy per day. It is assumed that the protein in a balanced diet should provide at least 10% of a person's energy. If the energy value of protein is 17 kJ g^{-1}, what is the minimum weight of protein which should be eaten each day?

2

ERRORS

There is a fundamental difference between counting and measuring; the former can always be done exactly, the latter never. If this seems odd to you, count the letters in the first sentence of this chapter—there are 101. Suppose that you wanted to know the mass of metal required in the type used to print these 101 letters. You weigh it and depending on the sensitivity of the balance obtain results "to the nearest 10 g, 1 g, 0.01 g..." But whatever the delicacy of the instrument there always remains some uncertainty, some limit to the accuracy of its measurements.

This is not the place for a discussion of the different sources of error, but for giving help in manipulating experimental data which include errors, such as 11.7 ± 0.3 m. Think about an experiment which is repeated several times by one person using the same instruments. It is very unlikely that the results will be the same each time, but we would expect them to be clustered round a central value. The more inaccurate the measurements the wider the range of values expected; and in reporting, the experimenter has to convey to the reader both the result and his or her confidence in it. This is done by writing, for instance, 11.7 ± 0.3 m, which states in short that you can be reasonably confident that if the experiment were repeated the result would lie between 11.4 m and 12.0 m. This being so it makes sense to quote results to 3 significant figures in this case and to keep at least 3 sig. fig. in any calculations concerning that quantity. Note that 14 ± 0.2 kg would be written as 14.0 ± 0.2 kg to show that 3 significant figures and not 2 are being retained.

There are mathematical techniques which can be applied to define error limits and to give precision to the "reasonable confidence" glibly mentioned in the last paragraph, and these will be touched on later in this book. However, a student usually has only a short time in which to complete an experiment and is unlikely to obtain more than 3 values for a particular measurement, and then elaborate methods are out of place. The variation between different members of a population being examined in biological field work is generally much greater than the error in the actual measurements of individuals, so that the techniques described here are used in chemistry or physics much more than in biology.

COMBINATION OF ERRORS

We can derive simple rules for combining errors if we make the assumption that $A = \alpha \pm \Delta\alpha$ means that the maximum value taken by A is $\alpha + \Delta\alpha$ and the minimum value is $\alpha - \Delta\alpha$. ($\Delta\alpha$ is read as delta alpha, the Greek letter Δ being commonly used to denote a small increment in a variable, in this case α). This is a simplification, replacing our "reasonable confidence" by certainty, but we make it, knowing what we are doing, since experience shows it to be sensible in the laboratory. Throughout this section we shall consider $A = 15.0 \pm 0.3$ and $B = 6.4 \pm 0.2$, forming in turn $A + B$, $A - B$, $A \times B$ and A/B.

ADDITION AND SUBTRACTION

Maximum value of $A = 15.3$, maximum value of $B = 6.6$; maximum value of $A + B = 15.3 + 6.6 = 21.9$.

Minimum value of $A = 14.7$, minimum value of $B = 6.0$; minimum value of $A + B = 14.7 + 6.2 = 20.9$, and all this can be expressed as

$$A + B = (15.0 + 6.4) \pm (0.3 + 0.2)$$

The maximum value of $A - B = 15.3 - 6.2 = 9.1$, the minimum value of $A - B = 14.7 - 6.6 = 8.1$, and this we could write as

$$A - B = (15.0 - 6.4) \pm (0.3 + 0.2).$$

Thus when adding or subtracting A and B, estimate the error in the new quantity by adding $\Delta\alpha$ and $\Delta\beta$. In practice it is unlikely that the maximum error will occur simultaneously in A and in B, but it might. When several quantities are combined the error calculated in this simple way is likely to be an overestimate and further thought must be given to the result.

PERCENTAGE ERRORS

Before finding the errors in $A \times B$ and A/B we must express $\Delta\alpha$ as a percentage of α (that is, calculate $(\Delta\alpha/\alpha) \times 100$). In our example

$$\frac{\Delta\alpha}{\alpha} \times 100 = \frac{0.3}{15} \times 100$$

$$= 2\%$$

and we say that the percentage error in A is 2%. The percentage error in B is easily found to be 3.125%, approximately 3%; let us call these p_A and p_B respectively. At once an advantage of this form appears. A has been measured to 2 parts in 100, B only to 3 parts in 100. Money transactions, so often carried out to 2 decimal places, show this very clearly. A child who has 10 pence and loses one loses 10% of his wealth; the car dealer selling a car whose price he reckons is £2000.02 loses only 0.001% of the price if he sells at £2000, and yet his loss is twice that of the child. Obvious? Of course; but the illustration may serve to remind you of the difference between errors expressed in these two ways.

FRACTIONAL ERRORS

For A and B these are $\frac{3}{150} = \frac{1}{50}$ and $\frac{2}{64} = \frac{1}{32}$. The form already used, 1 part in 50, also occurs.

MULTIPLICATION AND DIVISION

We use this concept of percentage error to derive expressions for the errors in $A \times B$ and A/B, making the same assumption as before. If the mathematical development does not interest you, there is a summary of the method for multiplication and division at the end of this section.

The percentage error in A is p_A, given by $p_A = (\Delta\alpha/\alpha)\,100$, using the same notation as before; thus

$$\begin{aligned} \Delta\alpha &= p_A\alpha/100 \\ A \times B &= (\alpha \pm \Delta\alpha)(\beta \pm \Delta\beta) \\ &= (\alpha \pm p_A\alpha/100)(\beta \pm p_B\beta/100) \\ &= \alpha\beta\left\{1 \pm \left(\frac{p_A}{100} + \frac{p_B}{100}\right) + \frac{p_Ap_B}{10000}\right\}, \end{aligned}$$

$p_Ap_B/10^4$ is very small and may be neglected, so we have

$$A \times B \approx \alpha\beta\left\{1 \pm \left(\frac{p_A}{100} + \frac{p_B}{100}\right)\right\}.$$

It can be shown, using a similar argument, that

$$\frac{A}{B} \approx \frac{\alpha}{\beta}\left\{1 \pm \left(\frac{p_A}{100} + \frac{p_B}{100}\right)\right\}.$$

MANIPULATION OF ERRORS; SUMMARY

ADDITION AND SUBTRACTION

Add the errors.

MULTIPLICATION AND DIVISION

Use the percentage errors p_A and p_B in $A = \alpha \pm \Delta\alpha$ and in $B = \beta \pm \Delta\beta$, $p_A = 100\,(\Delta\alpha/\alpha)$ and $p_B = 100\,(\Delta\beta/\beta)$, and

$$A \times B \approx \alpha\beta\left\{1 \pm \left(\frac{p_A}{100} + \frac{p_A}{100}\right)\right\},$$

$$\frac{A}{B} \approx \frac{\alpha}{\beta}\left\{1 \pm \left(\frac{p_A}{100} + \frac{p_B}{100}\right)\right\}.$$

First calculate the percentage errors in A and B, then add these together. Next calculate $\alpha\beta$ (or α/β in division) and use the combined percentage error to find the required error in $A \times B$ or A/B.

In our example $p_A = 2\%$, $p_B = 3\%$, $p_A + p_B = 5\%$.

$$\begin{aligned} A \times B &= 15.0 \times 6.4\,(1 \pm 0.05) \\ &= 96.0 \pm 4.8. \end{aligned}$$

$$\begin{aligned} A/B &= (15.0/6.4)\,(1 \pm 0.05) \\ &= 2.34 \pm 0.12. \end{aligned}$$

MEAN

It sometimes happens that you have several measurements of the same quantity, and although the foregoing simplified theory suggests that the mean is no more reliable than the individual results, common sense tells us that is not so. So to progress further we have to abandon the simplified theory because it does not take into account the tendency of errors to compensate for each other. The mathematical analysis is beyond the scope of this chapter (we return to the problem again in chapter 11), so we can only give you an expression which yields a reasonable estimate of the error, $\Delta\bar{x}$, of the mean $\bar{x}$ of n results of an experiment, namely

$$\Delta\bar{x} \approx \frac{\Delta x}{\sqrt{n}}.$$

Thus if the mean $\bar{x}$ of 9 readings were found to be 23.0, the error in each reading being $\pm$ 0.6,

$$\Delta\bar{x} \approx \pm 0.6/\sqrt{9} = \pm 0.2,$$

and we record the result as

$$\bar{x} = 23.0 \pm 0.2.$$

It could also happen that you obtain a small number, say 3, of determinations of the same quantity, but have little idea of the errors involved. Then your best estimate of the quantity measured is the mean of your results, and all you can say of the errors is to give the maximum and minimum readings, the range of values obtained in fact.

SIGNIFICANT FIGURES IN CALCULATIONS

When performing a calculation there may be no mention of errors, but you have to decide on the accuracy with which to work and how many significant figures to quote in the answer. The problem was mentioned in example 1 of chapter 1. There

$$\begin{aligned} C &= 2 \times \pi \times 2.00\,\text{m} \\ &= 2 \times 3.14159 \times 2.00\,\text{m} \\ &= 12.56636\,\text{m}, \end{aligned}$$

using a calculator.

2.00 is a measurement, assumed to be accurate to 3 significant figures, and the 2, an integer, is there to double it (an integer is accurate to as many sig. fig. as you like). π can be quoted to any desired degree of accuracy and in this case 3.14 would be sensible and sufficient; a slide rule would give as accurate a result as the figures warrant, maintaining 3 significant figures in the working. A useful rule for the answer is to quote it to the lowest number of significant figures in the original sum (here 3) so we gave the result as

$$C = 12.6\,\text{m}.$$

This takes account of the fact that any error in the original measurement has been multiplied, making it more realistic to give this answer than the *apparently* sensible one of

$$C = 12.57.$$

Look, then, critically at the figures in your problem; choose suitable approximations for any constants and identify the integers. Then the degree of accuracy you can obtain is the lowest number of significant figures in the remaining quantities.

ERROR BARS AND GRAPHS

The uncertainty in a quantity being plotted can be represented by a line drawn on either side of the point on the graph. This line is called an error bar and is illustrated in figure 2.1 for the quantity, x, with uncertainty or error, Δx.

Figure 2.2 shows the error bars for uncertainties in both x and y, namely Δx and Δy.

Since each point on the error bar is a probable value for the quantity, error bars act as a useful guide in the drawing of the "least bad" line through the points of a graph.

Fig. 2.1.

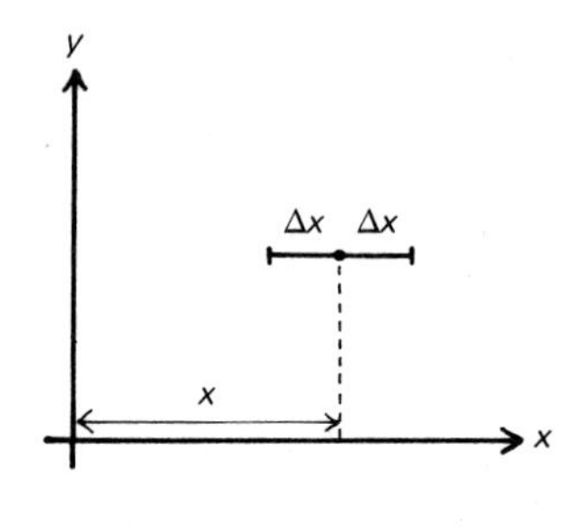

Fig. 2.2.

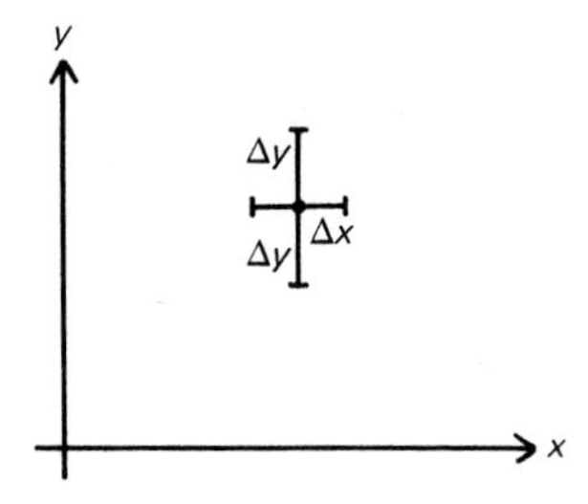

3

GROWTH, DECAY AND LOGARITHMS

Growth and decay are universal phenomena and the words are rich in associations for us. We think of physical growth, of growth of knowledge or of skills, sometimes with measurement in mind and sometimes without. Decay and decline are linked, or the word decay may immediately bring the compost heap to mind. Here however we are concerned with measurements and calculations and with natural laws; the "growth and decay" of the chapter heading refer to specific patterns of increase and decrease. Three examples illustrate this.

Fig. 3.1.

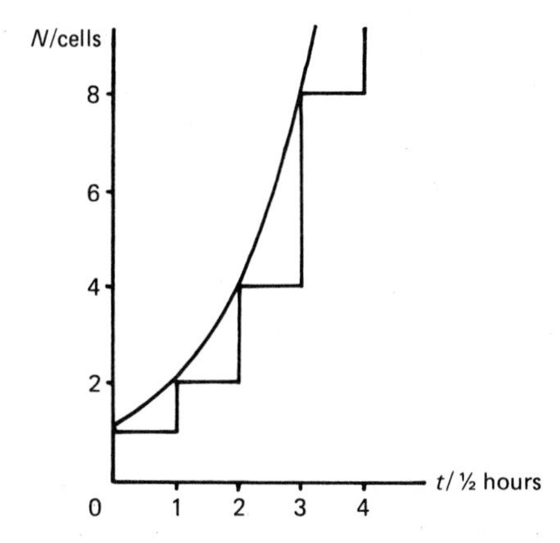

Example 1

GROWTH

A single bacterial cell divides after a period of $t = 30$ minutes; after another 30 minutes the 2 cells become 4 and so on, thus doubling the number of cells, N, every half-hour until other factors alter the pattern. Figure 3.1 shows this graphically.

For simplicity we have assumed that every cell has taken exactly 30 minutes to divide, but as the cells must be genetically similar this is a good first approximation to what actually happens. Expressed mathematically

$$N = 2^0 \qquad 0 \leqslant t < 1$$
$$N = 2^1 \qquad 1 \leqslant t < 2$$
$$N = 2^2 \qquad 2 \leqslant t < 3$$

and so on, or more briefly $N = 2^t$, $t = 0, 1, 2, \ldots$ units of 30 minutes if you remember that N remains at $2^3 = 8$ from $t = 3$ until the clock has passed another half-hour.

The same pattern of growth emerges when money is invested and allowed to accumulate.

Fig. 3.2.

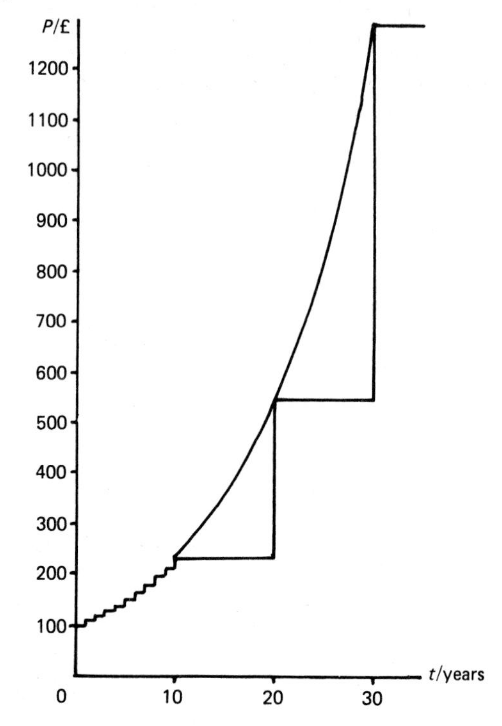

Example 2

GROWTH

£100 is invested at 9% per annum compound interest which is added on 1st January each year. Investigate the pattern of growth over a period of 30 years.

No approximation is involved here; the interest is not there on 31st December and is there on 1st January, after which no change occurs for one year. Figure 3.2 shows the growth curve.

The calculations are as follows:

Year 0, $P = £100$.

$$\text{Year 1, } P = £\left(100 + \frac{9}{100} \times 100\right) = £100\,(1 + 0.09)$$

$$= £100 \times 1.09.$$

$$\text{Year 2, } P = £(100 \times 1.09) \times 1.09 = £100 \times (1.09)^2,$$

and the tth year P is given by $P = £100 \times (1.09)^t$.

This is shown in detail in figure 3.2 for the first 10 years, but after that a smooth curve is drawn through a few calculated values. This is an approximation, but is easily understood as such and makes mathematical treatment much simpler.

Example 3

DECAY
At time $t = 0$ years, a cask is full of Glen Galloway whisky, a single matured malt. At time $t = 1$ year, half the quantity is drawn off and the cask filled up with a different whisky which we will call Whisky A. The cask, though once more full, contains $\frac{1}{2}$ Glen Galloway. At one year intervals the process is repeated and we record this in the table below, showing it graphically in figure 3.3.

t/years	0	1	2	3	...	t
E/casks	1	$\frac{1}{2}$	$\frac{1}{4}$	$\frac{1}{8}$	...	$(\frac{1}{2})^t$

This we express as

$$E = (\tfrac{1}{2})^t,$$

or $$E = 2^{-t}, \quad t = 0, 1, 2, \ldots \text{years},$$

remembering again, as in example 1 that E remains constant for one unit of time after each change. Notice that the Glen Galloway content diminishes each year but will never reach zero in theory. In practice the amount remaining soon ceases to be measurable.

In these rather fanciful examples the change is discontinuous, but the corresponding continuous curves are shown in figure 3.1 and figure 3.3. If

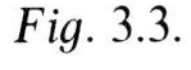
Fig. 3.3.

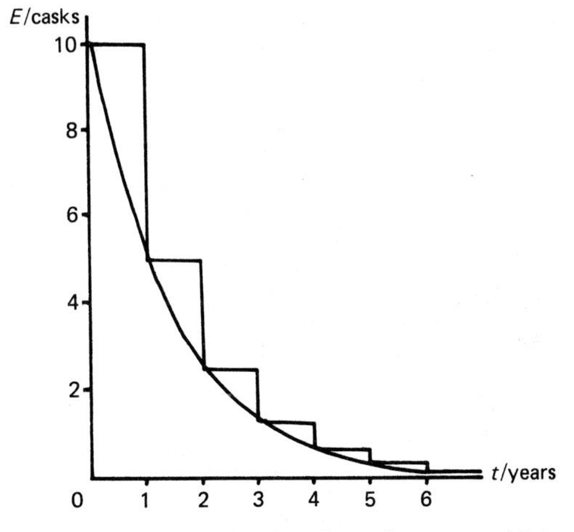

the function varied continuously the three laws would be stated as $N = 2^t$ for the cell growth, $E = 2^{-t}$ for the whisky, $P = £100 \times (1.09)^t$ for the investment.

EXPONENTIAL LAW
A variation is called exponential if it can be expressed either in the form
$y = ka^x$ *growth* (the cell growth and investment)

or $y = ka^{-x}$ *decay* (whisky) where k, a are constants and both >0.
We summarise the special characteristics of $y = a^x$, leaving further discussion to chapter 10, and you are advised to refer to figure 3.4 while reading the following notes. Domain, range and function are discussed in chapter 5.
1. While x can take any real value, negative, positive or zero, y is always positive and never zero. In other words the domain is the real numbers, the range real numbers >0. From the graphs you will see however that y approaches more and more closely to 0, as x becomes smaller.
2. $a^0 = 1$ whatever the value of a.
3. a is any positive constant $(a > 0)$ and all curves $y = a^x$ have the same essential shape. In figure 3.4 the three graphs are

$$y = 2^x, \quad y = e^x \ (e = 2.718\ldots), \quad y = 10^x.$$

4. If $y = a^x$ there is one and only one value of y corresponding to each value of x. In other words if $f: x \to a^x$ both f and f^{-1} are functions.
5. It follows from (4) that any positive real number can be expressed in the form a^x. Thus from figure 3.4 you can see that $8 = 2^3 \approx e^{2.07} \approx 10^{0.903}$; x, 3, 2.07 and 0.9 are called the exponents of a, 2, e and 10 respectively.

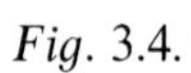
Fig. 3.4.

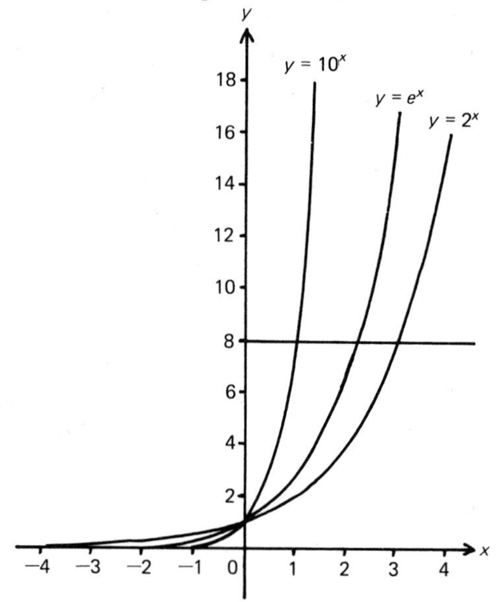

LOGARITHMS
The statement $2^3 = 8$ can also be expressed in the form

$$\log_2 8 = 3,$$

read as the log of 8 to the base 2 is 3 (log short for logarithm).

Thus $$e^{2.07} = 8 \Leftrightarrow \log_e 8 = 2.07$$

and $$10^{0.903} = 8 \Leftrightarrow \log_{10} 8 = 0.903,$$

where $\Leftrightarrow$ means if and only if, i.e. that one statement implies the other. In general,

$$a^x = y \Leftrightarrow \log_a y = x,$$

and the essential property of logarithms may be stated as

$$\log (p \times q) = \log p + \log q$$

(the base has been left out deliberately to emphasise the fundamental nature of this relation which is true regardless of the value of the base).

Whatever the base, $\log 1 = 0$. Bases, which must be positive, are usually

Fig. 3.5.

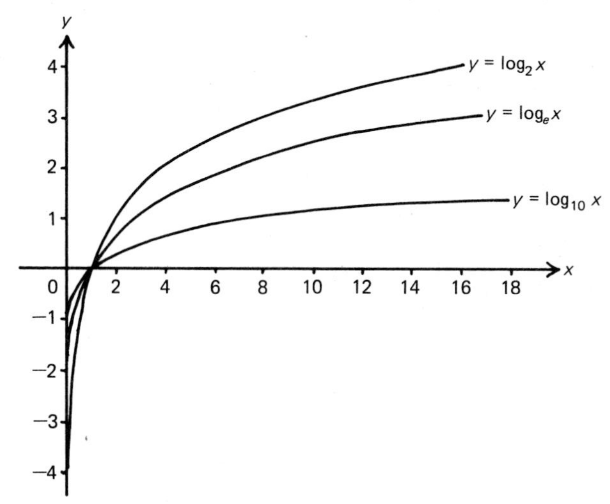

chosen to be greater than 1, and in that case when $x > 1$, $\log x > 0$,

when $x = 1$, $\log x = 0$, $0 < x < 1$, $\log x < 0$.

Refer to figure 3.5 to see this graphically. There the graphs of $y = \log_2 x$, $y = \log_e x$ and $y = \log_{10} x$ are plotted, corresponding to the exponentials shown in figure 3.4.

Note that $\log_a a = 1$; obvious if you think about it, but worth stating.

Exercise A

The problems which follow are given to aid in understanding the mathematical work of the chapter so far and are not direct applications from the sciences.

1. Write down the values of 2^{-3}, 3^4, 10^0, 6^{-2}, $(\frac{1}{2})^4$.
2. Draw the graph, from $x = -3$ to $x = +5$, of $y = 3^x$.
3. Express (a) 27 as 3^x, (b) 64 as 2^x, (c) 64 as 4^x, (d) 64 as 8^x, (e) 625 as 5^x.
4. Write down (a) $\log_3 27$, (b) $\log_2 64$, (c) $\log_4 64$, (d) $\log_8 64$, (e) $\log_5 625$.
5. For this question use the graphs of figure 3.4. Express each of the following numbers in three ways, namely as 2^x, e^x and 10^x; 2, 3, 6.5, 9, 0.8, 0.3.
6. Using the fact that $x = \log_a y \Leftrightarrow y = a^x$, rewrite the following statements: (a) $\log_3 9 = 2$, (b) $50 = 10^{1.699}$, (c) $\log_5 0.2 = -1$.

COMMON AND NATURAL LOGARITHMS

When calculation is all-important the obvious choice of base for logarithms is 10; such are the logarithms called Common Logarithms in books of tables, summoned in the calculator by pressing the key marked "log". However, in mathematical contexts, particularly in calculus, the base e occurs naturally and in books of tables these are to be found entitled Natural or Napierian Logarithms; the last name commemorates their discoverer, Baron Napier of Merchistoun (1550–1617). So frequently is this base used that there is a special notation for logs to the base e. We may write $\log_e x$ or simply $\ln x$, and the natural log key on calculators is very often labelled "ln".

Example 4
The law of growth is often expressed in the form

$$\frac{N}{N_0} = e^t,$$

where N is the number of the population at time t; N_0 is the number of the population when $t = 0$; t is the time, measured in appropriate units.

Find t as a function of N and N_0, and hence calculate t for a twofold, and for a 20-fold increase in the population.

$$\frac{N}{N_0} = e^t \Leftrightarrow t = \ln \frac{N}{N_0}.$$

For $\quad N = 2N_0, \quad t = \ln \dfrac{N}{N_0} = \ln 2 = 0.6931$ units of time.

For $\quad N = 20N_0, \quad t = \ln \dfrac{N}{N_0} = \ln 20 = 2.9957$ units of time.

CHANGE OF BASE
You can see from figure 3.5 that a given number x has a logarithm to any base (here there are three, 10, e, 2); sometimes you need to change from one base to another. This is easy between base 10 and base e if you have an electronic calculator (though it may take a few minutes to work out the method) but for reference we give the general case, changing from base a to base b.

$$y = \log_a x \Leftrightarrow a^y = x;$$

taking logs to the base b of both sides we have

$$y \log_b a = \log_b x.$$

Thus to change from base a to base b, multiply the original logarithm y by $\log_b a$.

Example 5
Use logarithms to the base 10 to calculate the time taken to double the population if

$$\frac{N}{N_0} = e^t$$

as in example 4.

$$\frac{N}{N_0} = e^t; \quad \text{but} \quad \frac{N}{N_0} = 2 \Rightarrow 2 = e^t.$$

Taking logs to the base 10 of both sides of this equation we have

$$\log_{10}2 = t\,log_{10}e,$$

$$t = \frac{\log_{10}2}{\log_{10}e} = \frac{0.3010}{0.4343}$$

and $$t = 0.6931 \text{ units.}$$

Exercise B provides practice not only in converting logarithms from base 10 to base e but also in the use of tables of Napierian logarithms since the answers worked out for $\log_e y$ can be directly checked using these tables when y itself has been found.

Exercise B
Copy and complete the following table

$\log_{10}(y)$	0.3010	1.3010	2.0000	$\bar{1}.9031$	-0.3010
$\log_e(y)$	0.6931				
y	2				

USES OF LOGARITHMS

THE QUANTITY pH

The degree of acidity of an aqueous solution varies with the hydrogen ion concentration. A convenient quantity used to measure the wide range of $[H_3O^+(aq)]$ is termed the pH of a solution. In dilute aqueous solutions, less than 0.1 mol dm^{-3}, pHhas a practical definition such that if

$$[H_3O^+(aq)] = 10^{-x}\,\text{mol}\,\text{dm}^{-3}, \quad \text{then} \quad p\text{H} = x,$$

or rather more formally

$$p\text{H} = -\log_{10}[[H_3O^+(aq)]/\text{mol}\,\text{dm}^{-3}].$$

Clearly acidic solutions will have pH values which are low or negative.

THE p NOTATION

This simply means "minus $\log_{10}$ of" and it can be applied to other quantities. Thus

$$p\text{OH} = -\log_{10}[[OH^-(aq)]/\text{mol}\,\text{dm}^{-3}].$$

Example 5
Calculate the pH of a solution containing $[H_3O^+(aq)] = 4.0 \times 10^{-3}$ M.*

* M represents mol dm^{-3}.

Method (a)

$$[H_3O^+(aq)]/\text{mol dm}^{-3} = 4.0 \times 10^{-3}$$
$$= 10^{0.60} \times 10^{-3} \quad (\log_{10}4 \simeq 0.60)$$
$$= 10^{-2.40}$$
$$\therefore \quad pH = 2.40.$$

Method (b)

$$pH = -\log_{10}[[H_3O^+(aq)]/\text{mol dm}^{-3}]$$
$$= -\log_{10}(4.0 \times 10^{-3})$$
$$= -(\bar{3}.60) = -(-3 + 0.6)$$
$$= +3.0 - 0.60$$
$$\therefore \quad pH = 2.40.$$

Example 6
If $[OH^-(aq)] = 1.0\,\text{M}$, calculate pOH.

Method (a)

$$[OH^-(aq)]/\text{mol dm}^{-3} = 1.0 = 10^0$$
$$\therefore \quad pOH = 0.$$

Method (b)

$$pOH = -\log_{10}[[OH^-(aq)]/\text{mol dm}^{-3}] = -\log_{10} 1.0 = 0.$$

Exercise C
Calculate the pH of solutions whose $[H_3O^+(aq)]$ are as follows:
(a) 10^{-4} M, (b) 5.7×10^{-6} M, (c) 0.002 M, (d) 10^{-14} M, (e) 0.03 M, (f) 2.0×10^{-2} M, (g) 7.0×10^{-5} M, (h) 9.1×10^{-4} M, (i) 4.4×10^{-2} M, (j) 0.16 M.

THE SELF-IONIZATION CONSTANT FOR WATER
This is defined as

$$K_w = [H_3O^+(aq)][OH^-(aq)].$$
$$\therefore \quad \log_{10} K_w = \log_{10}[H_3O^+(aq)] + \log_{10}[OH^-(aq)]$$
$$\therefore \quad -pK_w = -pH - pOH$$
$$\text{or} \quad pK_w = pH + pOH.$$

Now $$K_w = 1 \times 10^{-14}\,\text{mol}^2\,\text{dm}^{-6} \text{ at } 25°\text{C},$$
$$\therefore \quad pK_w = -\log_{10} K_w = -\log_{10}(10^{-14}) = 14.$$

Hence $$14 = pH + pOH \text{ at } 25°\text{C},$$

and this is a very useful relationship. Clearly in an exactly neutral solution $pH = pOH = 7$.

Example 7
A solution has a pH of 4.3. Calculate its $[H_3O^+(aq)]$ and $[OH^-(aq)]$ at 25°C.

Method (a)

$$pH = 4.3$$
$$\therefore \quad [H_3O^+(aq)] = 10^{-4.3}\,\text{mol dm}^{-3}.$$

However, it is normal practice to express a chemical concentration in standard form so we write

$$[H_3O^+(aq)] = 10^{-4.3} = 10^{0.7} \times 10^{-5} = 5 \times 10^{-5}\,M.$$

Now $[H_3O^+(aq)][OH^-(aq)] = 10^{-14}\,mol^2\,dm^{-6}$ in every aqueous solution at 25°C.

$$\therefore \quad [OH^-(aq)] = \frac{10^{-14}}{[H_3O^+(aq)]} = \frac{10^{-14}}{10^{-4.3}} = 10^{-9.7}$$

$$= 10^{0.3} \times 10^{-10} = 2 \times 10^{-10}\,M.$$

Method (b)

$$pH = 4.3 = -\log_{10}[[H_3O^+(aq)]/mol\,dm^{-3}]$$
$$\therefore \quad -4.3 = \log_{10}[[H_3O^+(aq)]/mol\,dm^{-3}].$$

Using the bar notation

$$-4.3 = -5.0 + 0.7 = \bar{5}.7 = \log_{10}[[H_3O^+(aq)]/mol\,dm^{-3}];$$

take antilogs $\therefore \quad [H_3O^+(aq)] = 5 \times 10^{-5}\,M.$

We have seen that

$$pH + pOH = 14$$
$$\therefore \quad pOH = 14 - 4.3 = 9.7.$$

Hence

$$-\log_{10}[[OH^-(aq)]/mol\,dm^{-3}] = 9.7$$
$$\therefore \quad -9.7 = -10 + 0.3 = \overline{10}.3 = \log_{10}[[OH^-(aq)]/mol\,dm^{-3}];$$

take antilogs $\therefore \quad [OH^-(aq)] = 2 \times 10^{-10}\,M.$

Exercise D
Calculate the $[H_3O^+(aq)]$ and $[OH^-(aq)]$ in solutions with the following pH values at 25°C;
(a) 12.0, (b) 0.1, (c) 5.7, (d) 8.6, (e) 2.3, (f) 10.6, (g) −1.0, (h) −2.9, (i) 4.4, (j) 3.7.

NUMBER OF IONS IN SOLUTION
The following calculation illustrates the importance of considering the number of ions in solution.

Example 8
Estimate the $[H_3O^+(aq)]$ and pH in an aqueous solution of 0.002 M $Ca(OH)_2$ at 25°C. Assume complete ionization.

$$Ca(OH)_2\,(aq) \rightarrow Ca^{2+}(aq) + 2OH^-(aq).$$

Notice that one mole of $Ca(OH)_2$(aq) gives two moles of OH^-(aq),

$$\therefore \quad [OH^-(aq)] = 2 \times 0.002 = 4 \times 10^{-3}\,M.$$

Method (a)

$$[H_3O^+(aq)] = \frac{10^{-14}}{[OH^-(aq)]} = \frac{10^{-14}}{4 \times 10^{-3}} = 2.5 \times 10^{-12}\,M$$

$$= 10^{0.4} \times 10^{-12} = 10^{-11.6}\,\text{M},$$

$$\therefore \quad pH = 11.6.$$

Method (b) Use the relationship $pK_w = pH + pOH = 14.$

Now $\quad [OH^-(aq)] = 4 \times 10^{-3}\,\text{M}$

$$= 10^{0.6} \times 10^{-3} = 10^{-2.4}\,\text{M},$$

$$\therefore \quad pOH = 2.4.$$

Hence $\quad pH = pK_w - pOH = 14 - 2.4 = 11.6.$

DECIBELS

In these days of supersonic bangs and discotheque amplifiers it is not unusual to hear scientists describe levels of sound in terms of decibels. It is found that just as the ear identifies an octave between notes when the frequency doubles, whatever the original frequency, so increases in loudness depend on the *ratio* of the sound intensities and not on the absolute difference in intensities.

Thus comparisons of intensity or power can be expressed in bels where

$$\text{number of bels} = \log_{10} P_2/P_1,$$

P_1 and P_2 being the intensities or powers to be compared. The use of bels is not confined to sound alone and they can refer to any source of energy, e.g. electricity.

In practice, the decibel, which is defined as one-tenth of a bel, is used and hence

$$\text{number of decibels} = 10 \log_{10} P_2/P_1.$$

The threshold of hearing, which is the least audible sound at a frequency of 1000 Hz, corresponds to an intensity, P_1, of 1.00×10^{-12} watt m^{-2}. If the intensity changes to 1.26×10^{-12} watt m^{-2} then the difference in intensities of sound in decibels is

$$10 \log_{10}\left(\frac{1.26 \times 10^{-12}}{1.00 \times 10^{-12}}\right) = 10 \log_{10} 1.26$$

$$= 1.00\,\text{dB}.$$

This is the minimum change of power which the ear is able to detect.

On the other hand, a jet plane on take off can produce a sound intensity level 120 dB above threshold; which the following calculation shows is one million million (10^{12}) times increase above threshold.

$$120 = 10 \log_{10}\left(\frac{P_2}{10^{-12}}\right)$$

$$\frac{120}{10} = \log_{10} \frac{P_2}{10^{-12}}$$

Taking antilogs,

$$10^{12} = \left(\frac{P_2}{10^{-12}}\right).$$

STELLAR MAGNITUDES

Another unit using a logarithmic scale is star magnitude in astronomy. Just as the ear identifies ratios of intensities rather than absolute differences, the eye does the same with brightness, and the ancient Greek scale of 1st, 2nd, 3rd, ... magnitude (1st being the brightest) fits the modern logarithmic definition.

An increase of one magnitude represents a decrease of brightness by a factor of 2.512. Since $\log_{10} 2.512 = 0.4$, a change of 5 magnitudes (say from 1st to 6th) gives a factor of 100. Thus a 6th magnitude star is 100 times fainter than a 1st magnitude star—the same meaning to us as to the ancient Greeks.

4

GRAPHS

THE GRAPH AS VISUAL DISPLAY

An important stage in an experiment is often to plot a graph. You want to discover whether two quantities are related in some way, and by varying one deliberately while observing and measuring the other, you hope to discover this relation. A graph displays this relation visually, as you see from figure 4.1. Notice the basic cycle of 8 elements which is more difficult to determine from a table: in other words a periodic relationship is observed.

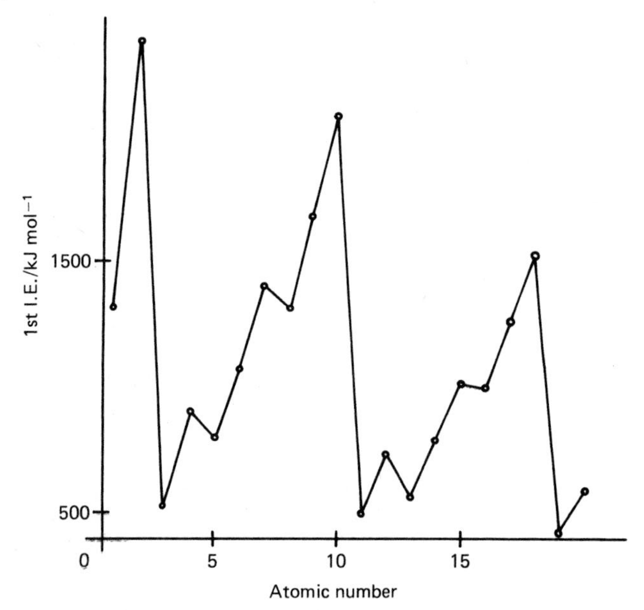

Fig. 4.1. *Graph of first ionisation energy* (1*st I.E.*) *against the atomic number of the element.*

Ability to interpret graphs comes only from a deeper understanding of their nature. Had you realised that graph paper is not essential? The distances of a point *P* from two fixed axes define its position; the grid lines on graph paper enable you to measure these distances quickly and accurately, but you can do without them. Two quantities are needed to fix a point on a surface, and two only: some well-known instances are latitude and longitude on the earth's surface, range and bearing, grid reference on a map.

CARTESIAN AND POLAR CO-ORDINATES

With the two axes at right angles the position of P is always given as $P(x, y)$, x units across the page (towards the right positive) and y units up the page (up positive). See figure 4.2. There is nothing sacred about drawing the axes at right angles, but the Cartesian system, named after René Descartes (1596–1650) who invented it, is extremely convenient and universally known. x is called the abscissa, or x-co-ordinate, y the ordinate or y-co-ordinate.

Another well-known method corresponds to range and bearing. Instead of references to two axes, a point O and a line OA are used. The polar co-ordinates of Q are then the distance from O, called r, and the angle measured anti-clockwise from OA, called θ (theta) and always written in the order (r, θ). (Figure 4.3.)

(x, y) and (r, θ) are ordered pairs of numbers—ordered because (3,5) is not the same point as (5,3). When you plot graphs of experimental results the points represent ordered pairs of related quantities such as (time, distance), (time, height of seedling), (volume of a gas, pressure). The shape which emerges should lead to the discovery of the relationship between them.

Fig. 4.2.

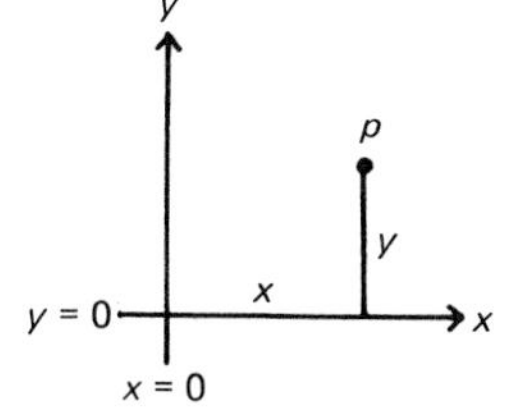

Fig. 4.3.

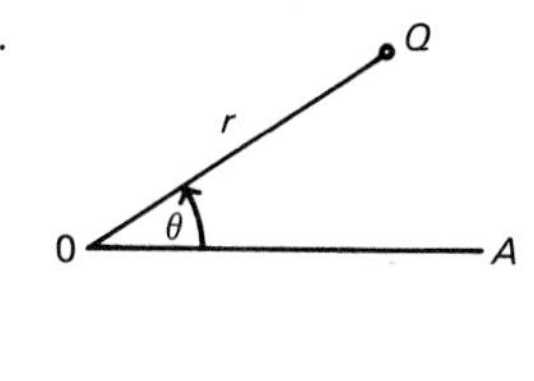

GRAPHS OF EXPERIMENTAL DATA; TECHNIQUES AND INTERPRETATION

Example 1

Below is a table giving the depressions e, of the free end of a cantilevered beam for different loads F at its free end. Plot a graph of e against F and from it determine the relation between e and F.

F/N	2.0	4.0	6.0	8.0	9.0
e/cm	6.5	13.1	19.7	26.1	29.6

The graph is shown in figure 4.4.

POSITION OF AXES

There are no negative values of e or F so the axes can cross at the bottom left-hand corner of the page which will be the origin O.

Fig. 4.4.

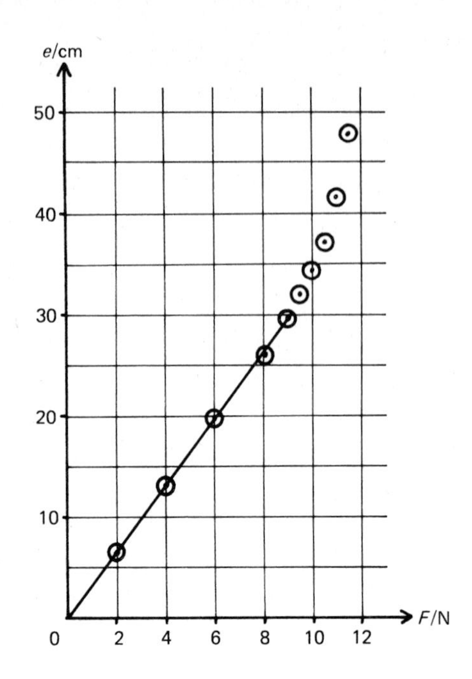

WHICH VARIABLE CORRESPONDS TO x?
Conventionally it is the independent variable, the quantity changed deliberately during the experiment (F in this case) and almost always recorded as the first line in a table of results. The phrase "e against F" gives the game away too, always "dependent against independent".

CHOICE OF SCALES
Use your paper to the fullest possible extent and choose your scales on this basis. Suitable scales would be

10 cm represent 10 N on the x (or F) axis,
10 cm represent 25 cm on the y (or e) axis.

Always avoid inconvenient scales if you can even if you cannot make full use of the page, because mistakes are more readily made when the scale is awkward.

LABELLING OF AXES
Number neatly to the left of the y axis and below the x axis in such a way that there is no uncertainty about the line to which the number refers. No other record of scale is needed when the quantity and the units used for it are clearly shown. There are two acceptable methods; e.g.
(i) e in cm, often used in physics, and
(ii) F/N, widely used in all scientific work.

PLOTTING POINTS
Use a sharp pencil (easily erased after the inevitable mistake) to record the

5 ordered pairs (2.0, 6.5), (4.0, 13.1), (6.0, 19.7), (8.0, 26.1), (9.0, 29.6) as points on the graph. Do not join up individual points.

DRAWING THE LINE

The points seem to lie close to a straight line so we fit one as well as we can. One method is to try to have the same number of points on each side of the line, moving the ruler, preferably transparent, until the best solution is found by eye. Sometimes there is a fixed point as in this example. where no deflection for no load means that (0,0), the origin, lies on the line. No experimental error is given for this data, but when errors are recorded, error bars are used, and the line must cut as many error bars as possible. Practice is the only thing that really helps in this rather difficult problem of finding the line of best fit by drawing. There are quantitative methods available, but they are not appropriate to this type of laboratory work.

FINDING THE RELATION BETWEEN THE VARIABLES e AND F

We assume that the straight line represents the relation which must therefore be linear. The line passes through the origin and so we know that e is directly proportional to F,

$$\text{i.e.}\quad e \propto F$$
$$\Leftrightarrow\quad e = kF \text{ where } k \text{ is a constant.}$$

In this case $k = 3.1$ and we have the required relation

$$e = 3.1\,F.$$

k is the gradient of the line (a concept discussed in detail later in this chapter) and since it is the rate of change of e with F, has units cm N^{-1} so that

$$k = 3.1 \text{ cm N}^{-1}.$$

The experiment was continued for forces greater than 9 N giving the results tabulated here.

Load F/N	9.5	10.0	10.5	11.0	11.5
Depression e/cm	32.0	34.4	37.3	41.6	47.9

Figure 4.4 shows that the relationship is no longer linear. Thus only over the domain $0 \leqslant F \leqslant 9.0$ N can we say that $e = 3.1\,F$.

RELIABILITY OF THIS RESULT

A rough guide is the closeness of the points to the line. In this experiment they lie close to it, so the result can be considered reliable.

Exercise A

In these examples you are asked to find, by graphical methods, the relationship between the variables tabulated. In each case plot a graph, then if the points lie approximately on a straight line go on to deduce the relation. If they do not appear to lie on a straight line, think why.

1. Using the following data draw graphs of °C against °F, and °C against K. Express these linear relationships mathematically, and comment on the importance of the intercepts in the latter graph.

K	73.2	123.2	173.2	223.2	273.2	323.2	373.2	423.2	473.2
°C	−200	−150	−100	−50	0	50	100	150	200
°F	−328	−238	−148	−58	32	122	212	302	392

2. The variation of the density of water with temperature is given. The relation is not linear; find the temperature for which the density is a maximum, labelling the appropriate point on the graph accordingly.

Temperature t/°C	0	2	4	6	8
Density/kg m^{-3}	999.88	999.96	1 000.00	999.96	999.89

Temperature t/°C	10	12	14
Density/kg m^{-3}	999.74	999.52	999.28

Note the very expanded scale necessary for density. Start the density scale at 999.0 so that your axes cross at (0, 999.0) and not at (0,0); this is often called a *false origin*.

3. The table shows the current I through a wire for different applied voltages V. The error in the ammeter reading is ± 1 mA and should be included. From a graph determine the relation between I and V.

Voltage V/V	0	0.5	1.5	2.5	−0.5	−1.5	−2.5
Current I/mA	0	8	22	37	−7	−23	−36

4. The variation of extension e, of a coiled metal spring with applied load, L, is given. Determine the relationship between e and L and the range over which it holds.

L/N	0	20	40	60	80	100	110
e/cm	0	6.61	13.3	19.8	26.2	33.1	38.4

5. For the following straight chain alkanes draw a graph of boiling point against the number of carbon atoms in the chain, using the tabulated data. How do you account for the shape of the graph?

Formula	CH_4	CH_3CH_3	$CH_3CH_2CH_3$	$CH_3(CH_2)_2CH_3$	$CH_3(CH_2)_3CH_3$
Boiling pt /K	111.66	184.52	231.08	272.65	309.22

6. Using the following information plot melting point against atomic number Z for the elements in Group 1A of the Periodic Table. Explain the shape of the graph.

Z	3	11	19	37	55
Element	Lithium	Sodium	Potassium	Rubidium	Caesium
Melting point/K	454	371	336	312	302

7. The vapour pressure of nitrogen P was recorded for varying temperatures, T.

Temperature T/°C	−210	−205	−200	−195	−190
Vapour pressure $P/10^3$ Pa	12.5	29.1	61.2	111.1	190.5

The graph is curved: choose scales carefully, and note that $T < -190$°C when deciding where to put the vertical axis.

8. Three requirements for photosynthesis are carbon dioxide, sunlight and warmth. In this experiment with the water plant *Elodea*, sunlight and warmth were kept constant, and the carbon dioxide concentration was varied by adding potassium hydrogencarbonate to the water. The rate of photosynthesis was measured by counting the number of oxygen bubbles rising to the surface per minute.

CO_2 concentration (arbitrary units)	1	5	10	25	50	100
Average no. of bubbles/min	6	10	14	23	57	119

Plot a graph and if possible fit a straight line. Will it be useful to derive the numerical relation between the variables? Do you think that the result for 25 units of carbon dioxide indicates a change in gradient at that point, or merely a reading with a large error?

9. A similar experiment was carried out, varying the light intensity by placing a lamp at different distances from the plant.

Distance d/cm	*Light intensity* $\propto 1/d^2$	*Average number of bubbles per minute*	
		At 20°C	*At* 30°C
Darkness	0	0	0
100	1×10^{-4}	14	16
36.5	7.5×10^{-4}	14	25

Using the same axes plot graphs for the results at the 2 different temperatures. (a) In certain thermochemical reactions the rate of reaction increases by a factor of about 2 for a rise of 10°C. Is this true of either light intensity? (b) Are you satisfied with the number of observations? Suggest 3 more light intensities which could have been used to give more information about the changing reaction rate at each temperature. Calculate the values of d which would give these intensities.

INTERPRETATION OF STRAIGHT LINE GRAPHS

In this section the properties of straight lines are investigated mathematically, without reference to experiment. It is easier to use x for the independent variable and y for the dependent variable throughout, and to follow the convention that other letters such as m, c, a, b, represent constants which do not vary in any one equation.

The equation of a straight line is always of the form

$$y = mx + c \tag{1}$$

and is called a linear equation (figure 4.5).

Fig. 4.5.

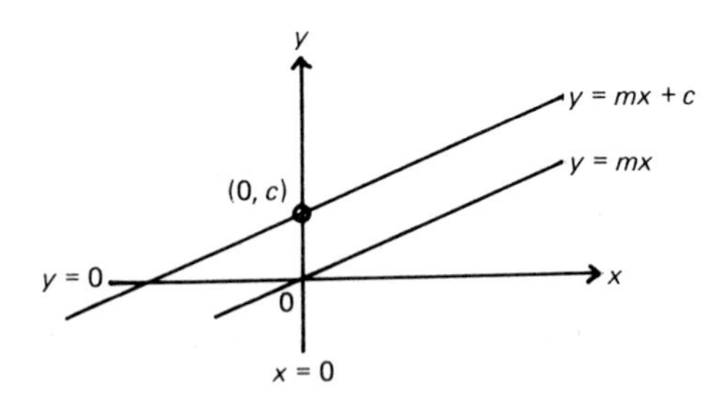

By putting $x = 0$ in (1) we find that $(0,c)$ lies on the line and c is called the y-**intercept**. In experimental work it is often referred to as the data zero, but whatever its name its magnitude is found by noting where the line (extrapolated if necessary) cuts the y axis. When $c = 0$, $y = mx$, and the line passes through the origin, $(0, 0)$.

GRADIENT

With the y-intercept fixed, the slope of the line is all that is needed to distinguish the line described by equation (1) from all others. The gradient

Fig. 4.6.

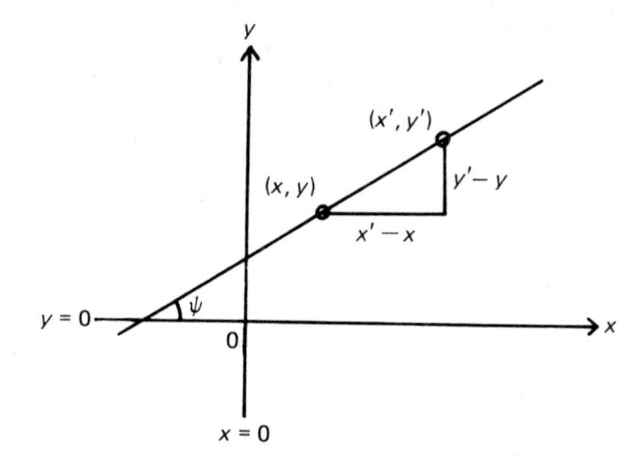

is defined mathematically (figure 4.6) by

$$\text{gradient} = \frac{y \text{ step}}{x \text{ step}}$$

$$= \frac{y' - y}{x' - x}$$

$$= m,$$

the constant multiplying x when the equation of the line is of the form of equation (1). (y step)/(x step) is only a mnemonic, but is a good way of remembering that to calculate gradient you choose a suitable and easy x-step and read off from the graph the corresponding y step. The gradient is sometimes called tan ψ (Greek letter psi) but only if the scales on x and y axes are the same is ψ the angle the line makes with the x axis.

Gradient is the same at every point on a straight line, and

gradient positive $\Rightarrow y$ increases as x increases,

gradient negative $\Rightarrow y$ decreases as x increases,

gradient $= 0 \Rightarrow y$ is constant ($\Rightarrow$ means "implies that").

Summing up, the gradient is the rate of change of y with x, and when the gradient represents the relation between physical quantities it must be quoted in terms of their units; in our example 1 gradient = 3.3 cm N^{-1} (note that the y-intercept was 0).

Example 2
Sketch the graph of $y = \frac{1}{2}x + 3$; find the y-intercept and the gradient.

$$y\text{-intercept} = 3, \quad \text{gradient} = \tfrac{1}{2}.$$

Sketch, by joining (0, 3) and (−6, 0) (figure 4.7).

SKETCHING A STRAIGHT LINE GRAPH QUICKLY
Method (a) Find y-intercept and draw a line through it with the required gradient: in example 2, line of gradient $\frac{1}{2}$ through (0,3).

Fig. 4.7.

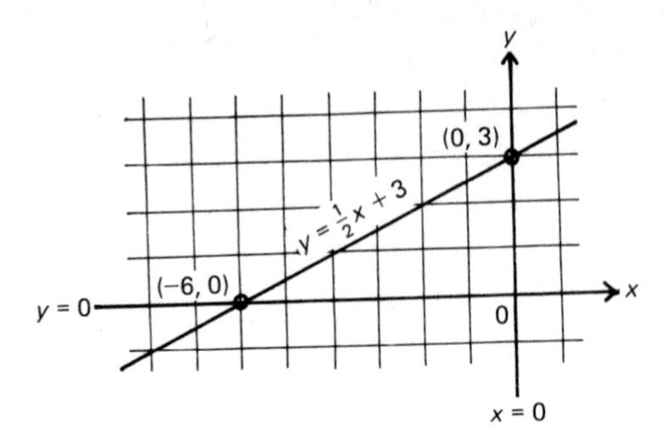

Method (b) Substituting $y = 0$ in the equation find the intercept on the x axis $(-c/m, 0)$ and join it to $(0, c)$: in example 2 join $(-6, 0)$ to $(0, 3)$.

Note that any other pair of points would do, and that a point and direction, or two points, define a straight line completely.

You can convert any linear equation in x and y to the form $y = mx + c$.

Example 3 illustrates the methods.

Fig. 4.8.

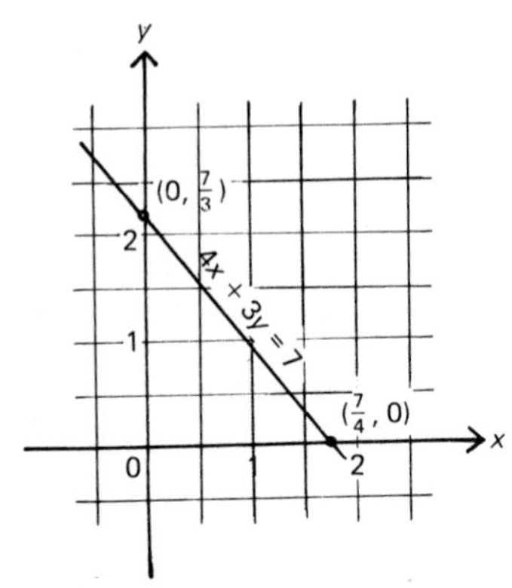

Put $4x + 3y = 7$ in the form $y = mx + c$.

$$4x + 3y = 7$$

divide throughout by 3

$$\frac{4}{3}x + y = \frac{7}{3},$$

and subtract $\frac{4}{3}x$ from both sides

$$y = \frac{7}{3} - \frac{4}{3}x$$

and rearrange;

$$y = \frac{-4}{3}x + \frac{7}{3}.$$

Gradient $= \dfrac{-4}{3}$ and y-intercept $= \dfrac{7}{3}$ (See figure 4.8).

Or use a flow diagram,

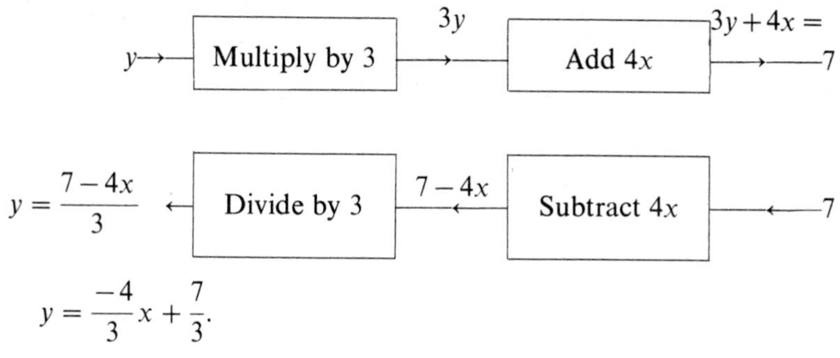

$$y = \frac{-4}{3}x + \frac{7}{3}.$$

Note that you are making y the subject of the formula here.

Exercise B

Where necessary put each of the following equations into the form $y = mx + c$. Find the y-intercept and the gradient for each, sketch the graphs and state whether y increases, decreases or remains constant with increasing x.

1. $y = 3x + 5$.
2. $y = 2x$.
3. $y = \frac{1}{3}x + 4$.
4. $7x - 3y = 5$.
5. $x + y = 4$.
6. $2x - y = 8$.
7. $y = 3$ for $x > 0$.

8. A fixed mass of gas is kept at constant pressure (atmospheric) and the variation of volume V with temperature T recorded. Plot a graph and by finding the y intercept and gradient find the relation between T and V. By extrapolation beyond the observed values find the temperature at which the volume tends to 0.

Temperature T/°C	21.0	45.0	68.0	83.0	100.0
Volume V/cm^3	1.14	1.25	1.32	1.38	1.45

9. The table shows the variation of electrical resistance with temperature. You are asked to plot a graph of R against t and from it to discover the gradient and the resistance when $t = 0$°C.

t/°C	20	39	61	85	100
R/Ω	0.60	0.62	0.69	0.75	0.78

10. In an experiment to study thermal conduction, a uniform, lagged metal bar, length 12 cm, had one end A maintained at 0°C and the other at -18°C. When conditions were steady the temperature t was read at several points x cm from A along the bar. Draw a graph and deduce the relation between t and x (the points $(0, 0)$ and $(12, -18)$ lie on the graph).

x/cm	3	6	9
t/°C	−2.5	−6.5	−10

NON-LINEAR RELATIONS

Fig. 4.9.

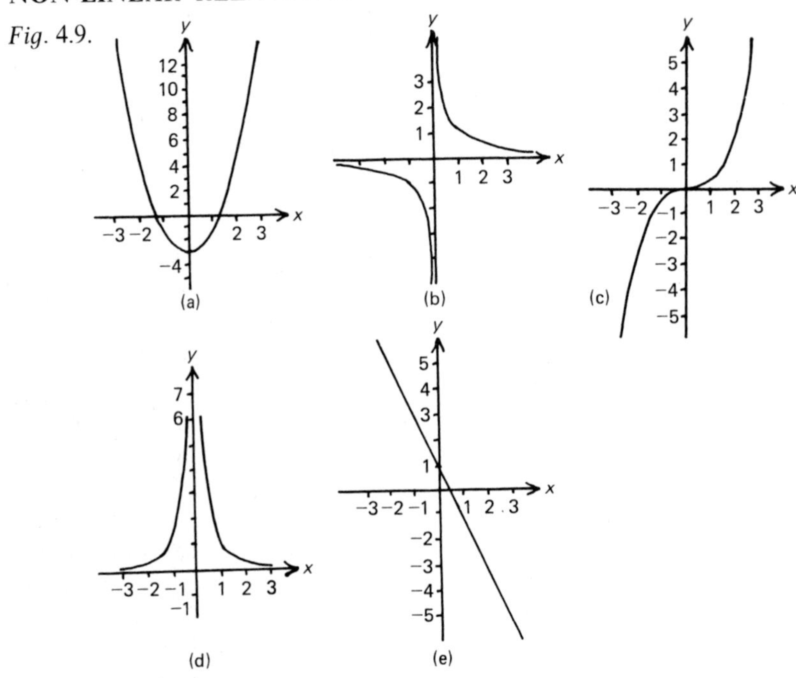

Five curves whose equations are of the form $y = ax^n + b$ are shown in figure 4.9 and two exponential curves, general equation $y = Pa^{kx}$, in figure 4.10. We summarise the information about these seven curves in the following table.

Figure 4.9, $y = ax^n + b$

	General equation	*Particular equation*	n	a	b	*Name*
(a)	$y = ax^2 + b$	$y = 2x^2 - 3$	2	2	−3	Parabola
(b)	$y = \frac{a}{x} + b$	$y = \frac{1}{x}$	−1	1	0	Rectangular hyperbola (Inverse law when $b=0$)
(c)	$y = ax^3 + b$	$y = \frac{x^3}{3}$	3	$\frac{1}{3}$	0	Cubic
(d)	$y = \frac{a}{x^2} + b$	$y = \frac{1}{x^2}$	−2	1	0	Inverse square law when b = 0
(e)	$y = ax + b$	$y = -2x + 1$	1	−2	1	Straight line—a special case of $y=ax^n+b$ when $n=1$.

Figure 4.10, $y = Pa^{kx}$

		P	a	k	
(a)	$y = 2^x$	1	2	1	Exponential growth
(b)	$y = 10 \times 3^{-x}$	10	3	-1	Exponential decay

Fig. 4.10.

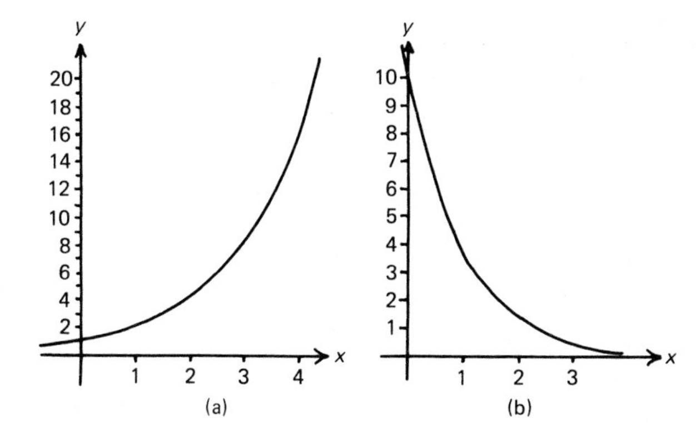

GRAPHICAL TECHNIQUES

A non-linear relation is investigated in detail in the next example. You are advised to work through it yourself, referring to the notes as you do so.

Example 4

A force is applied to a stationary trolley to which is attached a ticker tape. The distance of the trolley from its rest position is measured on the ticker tape for consecutive intervals of 0.2 seconds.

Time t/s	0	0.2	0.4	0.6	0.8	1.0
Distance s/cm	0	1.9	8.0	17.9	31.9	50.0

Use graphical methods to find the relation between distance travelled, s and time, t.

Fig. 4.11 (*a*)

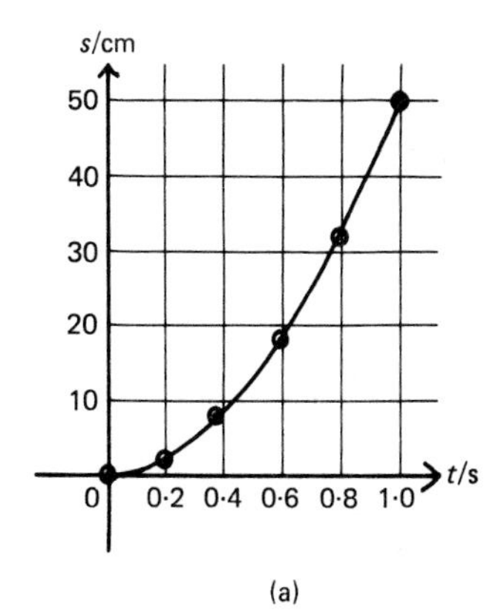

Fig. 4.11 (*b*)

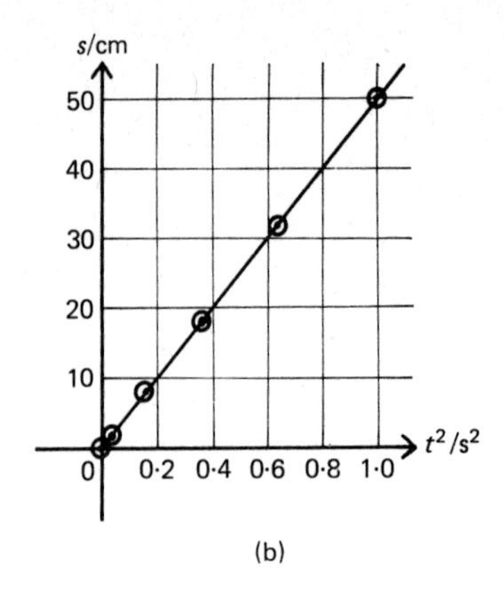

METHOD 1.

Plot a graph of s against t (figure 4.11(a)): the points (no error bars here) do not lie on a straight line, but near some other curve which must go through the origin. Choose suitable scales to use your graph paper effectively. Draw a smooth curve to fit the points, using a flexi-curve if you wish. Not only s, but the gradient, increases with t, and you may recognise the curve as a parabola where $b = 0$. This suggests that s may be proportional to t^2.

Calculate t^2 for each value of t, and plot a new graph, on new axes, of s against t^2 (figure 4.11(b)). The table is

t^2/s^2	0.04	0.16	0.36	0.64	1.00
s/cm	1.9	8.0	17.9	31.9	50.0

The points lie near a straight line through the origin so that we can say that s is proportional to t^2,

$$s \propto t^2$$

or

$$s = kt^2$$

where k is a constant.

To determine k we find the gradient of the graph;

$$k = 20/0.4 = 50\,\text{cm}\,\text{s}^{-2},$$

and the relation between s and t is

$$s = 50\,t^2.$$

METHOD 2, LOG-LOG GRAPH

Plot **log** s against **log** t (called a **log-log graph**): figure 4.12 shows this. Logarithms to the base 10 have been used in the following table, but logs to any base could be chosen.

t/s		0.2	0.4	0.6	0.8	1.0
log (t/s)	(i)	$\bar{1}.301$	$\bar{1}.602$	$\bar{1}.778$	$\bar{1}.903$	0
	(ii)	−0.699	−0.398	−0.222	−0.097	0
log (s/cm)		0.279	0.903	1.253	1.504	1.699

Calculators give version (ii) of log t/s, allowing you to plot the points (log t, log s) directly. Version (i) uses bar notation; in this case plot the points by finding −1 on the *log t* axis and moving in the positive direction from there. This is shown in figure 4.13 for $t = 0.4\,\text{s}$;

$$\log t = \log 0.4 = \bar{1}.602 = -1 + 0.602\,(= -0.398).$$

The points lie near a straight line with y-intercept = 1.69, gradient = 2. Thus

$$\begin{aligned}\log s &= 2\log t + 1.69;\ \text{but } 1.69 = \log 50 \text{ so that}\\ \log s &= 2\log t + \log 50, \text{ or}\\ \log s &= \log 50\,t^2,\\ s &= 50\,t^2.\end{aligned}$$

Fig. 4.12.

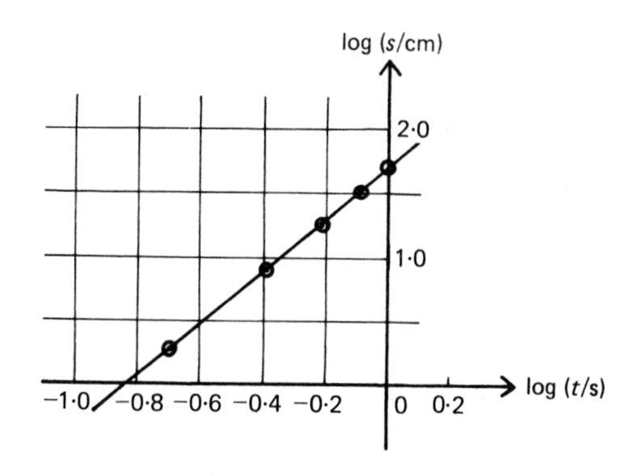

Fig. 4.13.

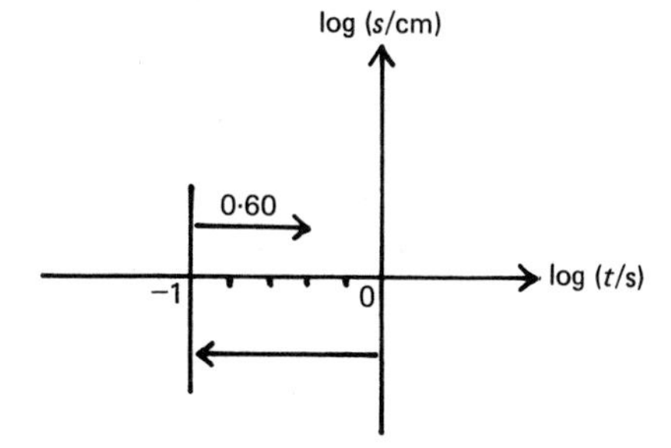

SUMMARY: FINDING NON-LINEAR RELATIONS GRAPHICALLY
From a table of data relating y to x you can plot a graph of y against x which gives you a curve whose shape you may recognize. You can also try to deduce the algebraic relation between y and x using one of the two following procedures.

METHOD 1
Assume that $y = ax^n + b$ and make an informed guess about n. Add to your table the values of x^n and plot a graph of y against x^n. If this proves to be a straight line as in figure 4.12 then $y = ax^n + b$, the gradient of the line being a and the y-intercept b. If the graph is not a straight line reject this hypothesis and start again with a new value of n.

METHOD 2
Plot *log y* against *log x* but only when there is no data zero. If $y = ax^n$ the graph will be a straight line whose equation is

$$\log y = \log a + n \log x.$$

Thus the gradient of the line gives n, the power of x (you do not have to guess its value as you do in method 1) and the y-intercept is log a. Logarithms to any base can be used.

The graph may be plotted on paper designed for log-log graphs if it is available.

Exercise C
In numbers 1 to 4 below you are asked to show, by graphical methods, the relationship between the variables tabulated, using the method outlined in the summary above. In numbers 5 to 7 there is a data zero so that it is not possible to use a log-log graph and help is given in deducing the relationship.

1. The variation of the length of a uniform column of trapped dry air with change in pressure (at constant temperature) is given. It can be assumed that the volume, V, of the air is directly proportional to the length of the column. Hence, find the relationship between the volume V of a gas and its pressure, P, at constant temperature (Boyle's law).

Pressure, $P/10^5$ Pa	1.15	1.21	1.30	1.40	1.51	1.62	1.71
Length of air column/cm	21.3	20.3	19.0	17.5	16.2	15.0	14.3

2. In an experiment to find the focal length of a converging lens, the distance, x, of the object from the focal point was measured along with the distance, y, of the image from the other focal point. Deduce the relationship between x and y.

x/cm	5	7	10	15	20	30
y/cm	46.8	33.0	23.4	15.6	11.7	7.7

If the focal length of the lens is actually 15.3 cm, what can you say further about the relationship?

3. The table shows the variation of resonant frequency, f, of a stretched string on a sonometer, with length, l, when the tension of this uniform string remained constant throughout the experiment. Find the relationship between f and l.

Frequency, f/Hz	256	341	384	427	480	512
Length, l/cm	40.5	30.0	27.2	24.4	21.6	20.2

4. Find the relationship between the electrical resistance, R, of a uniform wire, and its diameter, d.

Diameter, d/cm	0.60	0.99	1.5	2.2	3.1
Resistance, R/Ω	0.071	0.055	0.044	0.036	0.031

5. The table shows the corresponding values of the object distance, u, and image distance, v, for a converging lens of focal length, f.
   By drawing a graph of v against u you will see that the relationship between them is non-linear. It is, in fact,

$$\frac{1}{v}+\frac{1}{u}=\frac{1}{f}$$

Compare this with the general equation for a straight line, $y = mx + c$ and show that by plotting $1/v$ against $1/u$ a straight line is obtained with a negative gradient and equal intercepts. How can the focal length, f, be found from this graph?

u/cm	31.1	25.0	23.0	20.0	19.0
v/cm	29.5	38.9	45.0	62.5	72.7

6. A resonance tube, with end correction, e, resonates to the frequency, f, when the length of the air column is l. The table shows the corresponding values of l and f.
   The relationship is $l + e = v/(4f)$ where v is the speed of sound in air.
   How would you plot values of l and f to give a straight line graph? From this graph, determine v and e.

Frequency, f/Hz	256	341	384	427	480	512
Length of air column, l/cm	15.5	16.6	18.7	20.8	23.8	32.1

7. In an experiment to investigate the variation of intensity of γ radiation with distance from a small radioactive source, the count rate, c, was

measured as the distance, D, between the detector and source was varied. These values are tabulated.

Distance, D/cm	40	30	25	20	16	14
Count rate, c/counts per minute	48	124	157	263	432	540

To allow for the fact that the source and detector are of a finite size, an amount, x, had to be added to D to obtain the "true" source/detector separation.

The relationship is

$$c \propto \frac{1}{(D+x)^2} = \frac{1}{d^2}, \quad \text{where} \quad d = D + x.$$

By rearranging this relationship show that by plotting values of D against $1/\sqrt{c}$ a straight line graph is obtained from which the value of x can be determined.

8. The table shows the data obtained when several separate $100\,cm^3$ samples of $AgNO_3$ solution were reacted with different volumes of 0.1 M NaCl solutions.

Volume of 0.1 M NaCl/cm^3	20	40	60	80	100	120	140
Mass of AgCl precipitate/g	0.6	1.2	1.7	2.3	2.9	2.9	2.8

Draw a graph showing the relationship between the volume of NaCl solution and the resultant mass of AgCl precipitated. How do you account for the shape of the graph?

9. The Arrhenius equation states that the rate of a reaction has an exponential dependence, i.e. $k = Ae^{-E/RT}$, where A is a constant, E is the activation energy, R is the gas constant and T the absolute temperature. (a) Derive an expression for $\log_{10}k$. (b) How would you evaluate the energy of activation, E, by plotting a graph using $\log_{10}k$?

10. The vapour pressure P of a liquid is related to the absolute temperature T and to the enthalpy of vaporisation ΔH_{Vap} by the equation

$$2.303 \log_{10}(P/\text{mm Hg}) = \frac{-\Delta H_{\text{Vap}}}{RT} + \text{constant}.$$

Using the following information plot $\log_{10}(P/\text{mm Hg})$ against $1/T$.

Temperature/K	343	353	363	373
P/mm Hg	233.7	355.1	525.8	760.0

(a) Calculate ΔH_{Vap} from your graph.
(b) What is the boiling point of this liquid when the external pressure is 233.7 mm Hg?
Explain your answer.

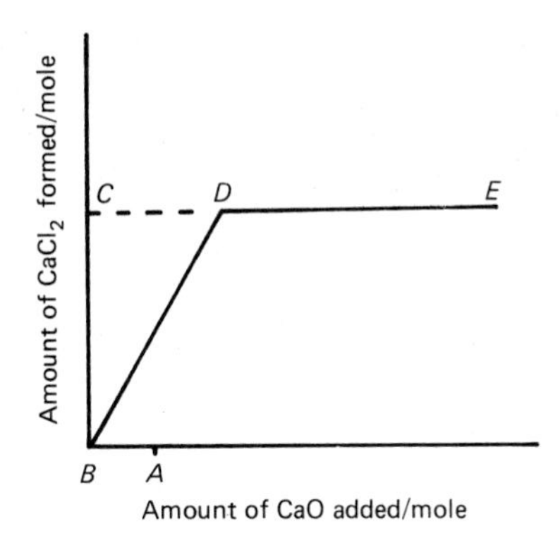

Fig. 4.14.

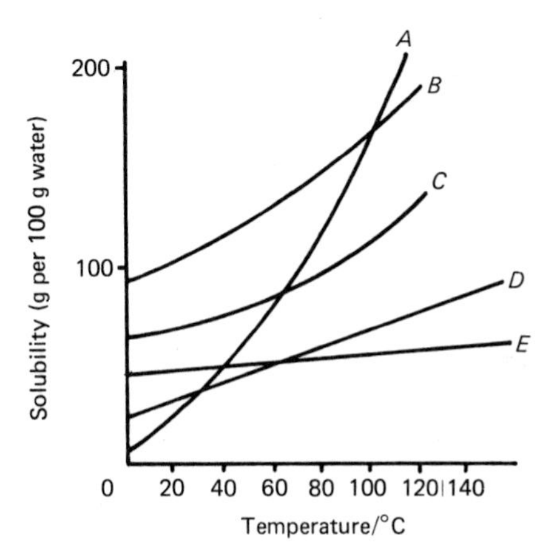

Fig. 4.15.

Exercise D

The following are miscellaneous examples provided to give practice in the interpretation of graphs.

1. Figure 4.14 represents data for the reaction

$$CaO(s) + 2HCl(aq) \rightarrow CaCl_2(aq) + H_2O(1).$$

What point on the graph corresponds to a 1:2 mole ratio of CaO to HCl: *A*, *B*, *C*, *D* or *E*?

Explain the significance of the horizontal section *DE*.

2. The solubility curves for substances *A*, *B*, *C*, *D* and *E* are shown in Figure 4.15.

(a) Which substance undergoes the greatest increase in solubility between 0°C and 60°C?

(b) Which substance has the highest solubility at 80°C?
(c) Which substance has the lowest solubility at 20°C?
(d) What is the difference in solubility between *C* and *E* at 100°C?

3. By reference to the phase diagram (figure 4.16) answer the following questions.
(a) At what temperature and pressure can all three phases of the substance exist simultaneously in equilibrium? What is the significance of the lines *AB* and *AC*?
(b) At what minimum temperature can the substance exist solely in the gaseous state if the pressure is 0.9 atm?
(c) If the substance is stored in a closed vessel at −80°C by what *factor* will its pressure rise if the temperature reaches −60°C?

Fig. 4.16.

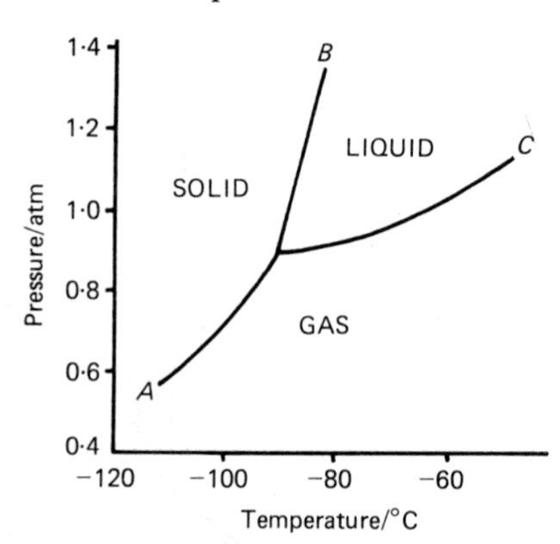

4. Draw a graph of volume of gas evolved against time for the reaction

$$Mg(s) + 2HCl(aq) \rightarrow MgCl_2(aq) + H_2(g).$$

Volume of hydrogen/cm^3	30	48	63	73	81	88	93	96	96	96	96	96
Time/min	0.5 *Z*	1.0	1.5	2.0 *Y*	2.5	3.0	3.5	4.0 *X*	4.5	5.0	5.5	6.0

(a) At which of the points marked *X*, *Y* and *Z* was the reaction fastest?
(b) What stage in the reaction is represented by the point *X* on the graph?
(c) After how many minutes had half the mass of magnesium reacted?
(d) If powdered Mg was used instead of Mg ribbon, what effect would this have on the rate?

5. If r is the distance of the electron from the nucleus, and ψ^2 is related to the probability of finding the electron at a particular point, comment on the significance of the graph (figure 4.17) of ψ^2 against r. Can you suggest a shape for this orbital?

6. (a) ^{14}N, ^{32}S, ^{55}Mn and ^{81}Br are all stable isotopes (nuclides). Copy and complete the table below, entering in it (i) the number of protons and (ii) the number of neutrons which a nucleus of each isotope contains.

	^{14}N	^{32}S	^{55}Mn	^{81}Br
No. of protons				
No. of neutrons				

(b) From the completed table plot a graph of the number of protons against the number of neutrons.
(c) Use your graph and any other data you need to suggest a likely atomic mass number for the stable isotope of silicon.

Fig. 4.17.

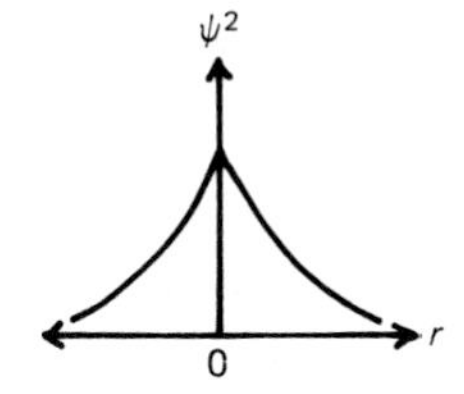

7. By inspection of the graph (figure 4.18) suggest which elements have the most stable nuclei.
8. The following results of measurements of pressure, P, and corresponding volume, V, were obtained from investigation of 1 mole of oxygen gas, and independently, of 1 mole of carbon dioxide gas at 273 K.

P/atm	0.2	0.5	0.8	1.0
Oxygen: $PV/\text{atm.dm}^3$	22.414	22.407	22.400	22.393
Carbon dioxide: $PV/\text{atm.dm}^3$	22.387	22.339	22.291	22.260

Fig. 4.18.

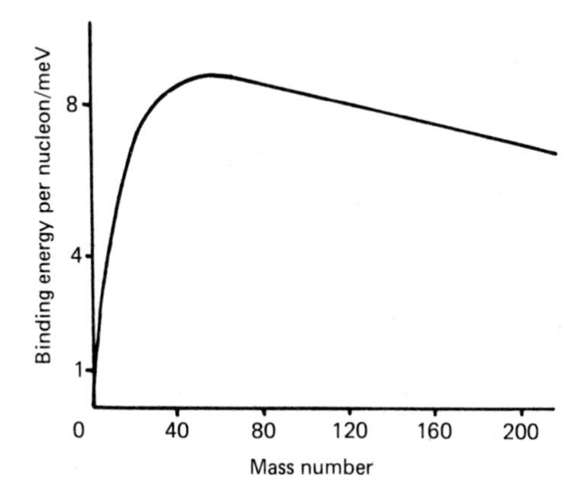

(i) Plot the values of PV against P for both oxygen and carbon dioxide.
(ii) Use your graph to show that, at very low pressures, both gases approach ideal behaviour.
(iii) Draw on your graph a dotted line to represent the behaviour of 1 mole of ideal gas.
(iv) Deduce from your graph the value of the gas constant R for an ideal gas, quoting the appropriate units.

9. The following data were obtained when investigating the variation of the volume of a gas with varying temperature at a fixed pressure. 0.434 g of gas was used at a pressure of 745 mm Hg.

Temperature/K	298	305	319	329	347
Volume/cm^3	339	346	362	374	394

(a) Plot the volume of gas, in cm^3, against temperature, in K.
(b) Use a graph to estimate the volume of the gas at 273K (V_0).
(c) Use the value of V_0 to find the relative molecular mass of the gas.
(d) Determine the gradient m of the graph in units of $cm^3\,K^{-1}$.
(e) Calculate the ratio V_0/m.
(f) State briefly the theoretical significance of the ratio V_0/m.

5

ALGEBRAIC TECHNIQUES

Scientific laws are expressed in words or in symbols, and calculations frequently involve the solution of equations. This chapter will revise the techniques you need; some methods and some of the language may be unfamiliar. By all means stick to the ways you know, but if you find you understand a new method better than the old, then change to it.

FUNCTIONS

The equation $y = 3x + 2$ defines a relation between x and y. Once the number x is chosen y is defined; thus for example 2 maps on to 8, -8 maps on to -22. This relation can be given a name, say f, and then we write either

$$f(x) = 3x+2, \text{ and } f(2) = 8, f(-8) = -22,$$

or $f:x \to 3x + 2$, read "f such that x maps on to $3x + 2$".

In this case for any one value of x there is only one value of $f(x)$, and the relation f is called a **function.** The **inverse relation** is often needed, written f^{-1} and read as "f inverse". If one value of $f^{-1}(x)$ corresponds to one value of x, $f^{-1}(x)$ is a function; flow diagrams provide a good method of finding f^{-1}. For the function $f:x \to 5x - 3$ and the inverse relation, $f^{-1}:x \to (x + 3)/5$ the flow diagram and its inverse are given below.

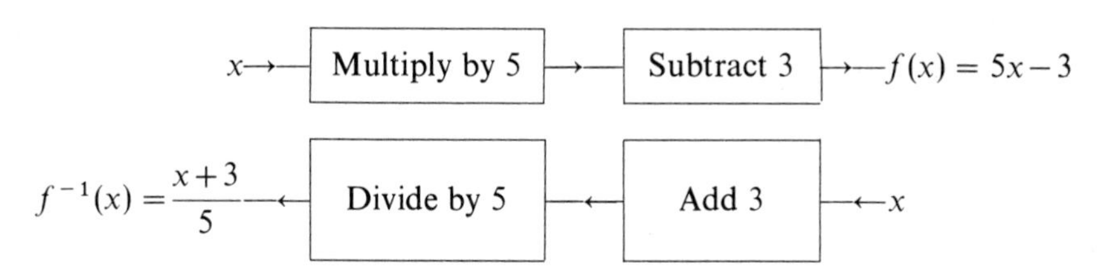

Here the inverse relation is a function.

The table which follows shows operations and their inverses in the form $f:x \to \ldots$ and in words.

Function	*In Words*	*Inverse Function*	*In Words*
$f:x \to x + b$	add b	$f^{-1}:x \to x - b$	subtract b
$f:x \to x - b$	subtract b	$f^{-1}:x \to x + b$	add b
$f:x \to bx$	multiply by b	$f^{-1}:x \to x/b$	divide by b
$f:x \to x/b$	divide by b	$f^{-1}:x \to bx$	multiply by b

In all of these both f and f^{-1} are functions. There are two special cases which, as you can check for yourselves, are self-inverse, that is, $f = f^{-1}$.

They are shown below.

$f{:}x \to b - x$ subtract from b
$f{:}x \to b/x$ divide into b

A common function where inverse is not a function is

$f{:}x \to x^2$ square x $f^{-1}{:}x \to \pm\sqrt{x}$ take square root of.

DOMAIN AND RANGE

The domain is the set of values of x over which the function is defined, the range the set of values which the function may take.

Example 1

For (i) $f(x) = 9x + 4$, (ii) $g(x) = 7x^2$, (iii) $h(x) = 2/x$, find the domain and range.

(i) $f{:}x \to 9x + 4$. x can take any real value, and so can $f(x)$, so domain and range are the same, all real numbers.

(ii) $g{:}x \to 7x^2$. x can take any real value, but $f(x)$ will always be positive unless $x = 0$. Thus the domain is all the real numbers but the range is the real numbers $\geqslant 0$.

(iii) $h{:}x \to 2/x$. x cannot take the value 0 so this must be excluded from the domain which is then the real numbers excluding 0. The range is the same as the domain.

CHANGING THE SUBJECT OF A FORMULA

Flow diagrams can help in manipulating formulae (and in solving equations) by showing clearly the order in which the necessary operations have to be done.

Example 2

Make T the subject of the formula $L = L_0(1 + \alpha T)$. The problem is solved in two ways, first using flow diagrams and then without them.

1.

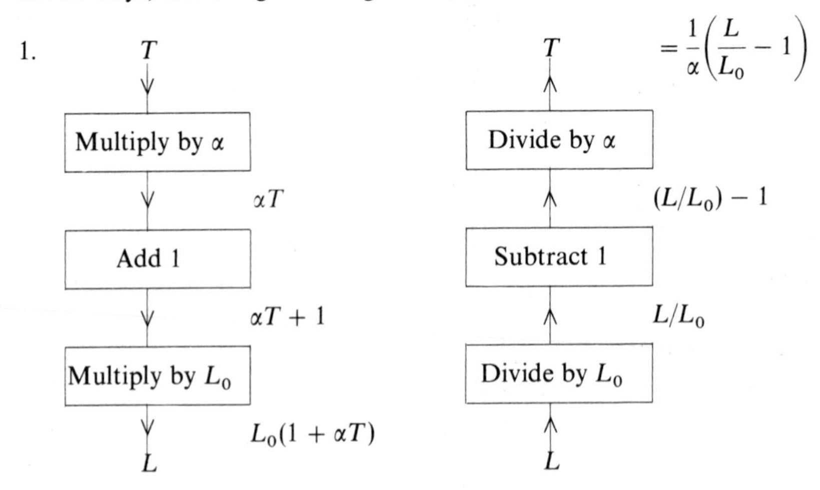

Thus $T = \frac{1}{\alpha}\left(\frac{L}{L_0} - 1\right) = \frac{L - L_0}{\alpha L_0}$.

Note that the inverse operations are applied in reverse order (when you go upstairs and then return down again the last step up is the first down).

2. $$L = L_0(1 + \alpha T)$$

$$\frac{L}{L_0} = 1 + \alpha T$$

$$\frac{L}{L_0} - 1 = \alpha T$$

$$T = \frac{1}{\alpha}\left(\frac{L}{L_0} - 1\right) = \frac{L - L_0}{\alpha L_0}.$$

Exercise A

In numbers 1 to 8 make the symbol named after the semi-colon the subject of the formula.

1. $\sigma = \frac{s}{\sqrt{n}}$; n.

2. $m = \frac{v}{u}$; u.

3. $\frac{1}{f_m} = \frac{d}{c}$; c.

4. $F = \frac{qQ}{4\pi\varepsilon_0 r^2}$; (i) ε_0 (ii) r.

5. $\frac{1}{R} = \frac{1}{R_1} + \frac{1}{R_2}$; R.

6. $\frac{(a+b)y}{2al} = \frac{4\pi ma}{c}$; c.

7. $\frac{P_1V_1}{T_1} = \frac{P_2V_2}{T_2}$; (i) T_1 (ii) V_1.

8. $\frac{1}{\lambda} = R_H\left(\frac{1}{n^2} - \frac{1}{m^2}\right)$; λ.

9. $t = 2\pi\sqrt{\frac{l}{g}}$; find l if $g = 9.81\ \text{ms}^{-2}$, $t = 1.1$ s.

10. $A = \pi R^2 - \pi r^2$; find A when $R = 10.41$ cm, $r = 10$ cm.

11. Show that van der Waals' equation

$$\left(P + \frac{a}{V^2}\right)(V - b) = RT,$$

(for 1 mole of gas) is the same as the following equation

$$PV^3 - (RT + Pb)V^2 + aV - ab = 0.$$

SIMULTANEOUS EQUATIONS

In general, two linear equations

$$ax + by = h$$
$$cx + dy = k,$$

are satisfied *simultaneously* by one ordered pair of numbers, say (x_1, y_1). Each equation describes a straight line, and the two lines intersect in a single point P whose co-ordinates are (x_1, y_1) as in figure 5.1. You can solve such equations graphically, or algebraically by elimination, substitution or by a matrix method. Before describing the latter technique we remind you about the identity matrix and inverse matrices.

Fig. 5.1.

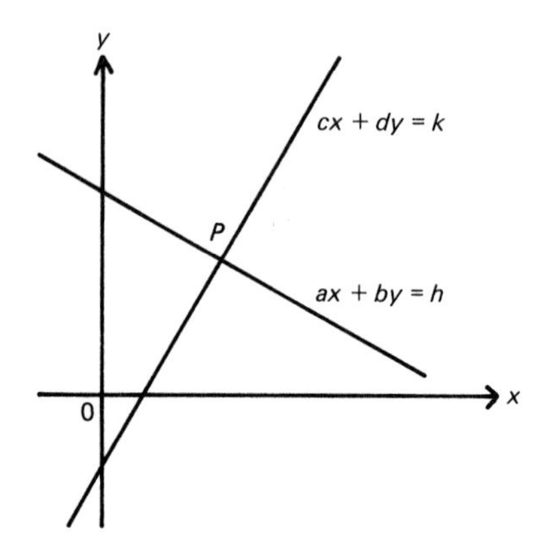

IDENTITY MATRIX; INVERSE OF 2×2 MATRIX

Let **I** be the **identity matrix.** Then for any 2×2 matrix **A**

$$\mathbf{IA} = \mathbf{AI} = \mathbf{A}, \quad \mathbf{I} = \begin{pmatrix} 1 & 0 \\ 0 & 1 \end{pmatrix}.$$

For a 2×2 matrix **A** the **inverse matrix** $\mathbf{A}^{-1}$ (read "**A** inverse") is defined by the relation

$$\mathbf{A}\mathbf{A}^{-1} = \mathbf{A}^{-1}\mathbf{A} = \mathbf{I}.$$

Check for yourself that if

$$\mathbf{A} = \begin{pmatrix} a & b \\ c & d \end{pmatrix}, \quad \mathbf{A}^{-1} = \frac{1}{ad - bc}\begin{pmatrix} d & -b \\ -c & a \end{pmatrix}, \quad ad - bc \neq 0.$$

When $ad - bc = 0$, $\mathbf{A}^{-1}$ does not exist.

$ad - bc = \det \mathbf{A}$ is called the **determinant** of **A**.

Example 3

If possible find $\mathbf{A}^{-1}$ and $\mathbf{B}^{-1}$, where

$$\mathbf{A} = \begin{pmatrix} 2 & 1 \\ 4 & 3 \end{pmatrix}, \quad \mathbf{B} = \begin{pmatrix} 5 & 7 \\ 10 & 14 \end{pmatrix}.$$

$\det \mathbf{A} = 6 - 4 = 2;$

$$\mathbf{A}^{-1} = \frac{1}{2}\begin{pmatrix} 3 & -1 \\ -4 & 2 \end{pmatrix} = \begin{pmatrix} \frac{3}{2} & -\frac{1}{2} \\ -2 & 1 \end{pmatrix}.$$

$\det \mathbf{B} = 70 - 70 = 0;$ $\mathbf{B}^{-1}$ does not exist.

SOLUTIONS USING MATRICES

Write the equations $ax + by = h$, $cx + dy = k$ in the form

$$\begin{pmatrix} a & b \\ c & d \end{pmatrix}\begin{pmatrix} x \\ y \end{pmatrix} = \begin{pmatrix} h \\ k \end{pmatrix}, \quad \text{or} \quad \mathbf{A}\begin{pmatrix} x \\ y \end{pmatrix} = \begin{pmatrix} h \\ k \end{pmatrix}.$$

Calculate $\det \mathbf{A}$; if $\det \mathbf{A} = 0$ there is no unique solution. For $\det \mathbf{A} \neq 0$, form $\mathbf{A}^{-1}$. Because $\mathbf{A}^{-1}\mathbf{A} = \mathbf{I}$,

$$\mathbf{A}^{-1}\mathbf{A}\begin{pmatrix} x \\ y \end{pmatrix} = \mathbf{A}^{-1}\begin{pmatrix} h \\ k \end{pmatrix},$$

$$\mathbf{I}\begin{pmatrix} x \\ y \end{pmatrix} = \begin{pmatrix} x \\ y \end{pmatrix} = \mathbf{A}^{-1}\begin{pmatrix} h \\ k \end{pmatrix}$$

which gives the solution.

This routine can always be used, but always begin by finding $\det \mathbf{A}$, thus saving work if no unique solution exists. Example 4 illustrates the three cases which can arise.

Example 4

Solve, if possible,

(i) $2x + 3y = 12$, $3x + 6y = 21$, (ii) $4x + 5y = 17$, $12x + 15y = 18$, (iii) $x + 2y = 6$, $3x + 6y = 18$.

(i)
$$\begin{pmatrix} 2 & 3 \\ 3 & 6 \end{pmatrix}\begin{pmatrix} x \\ y \end{pmatrix} = \begin{pmatrix} 12 \\ 21 \end{pmatrix} \quad \text{or} \quad \mathbf{A}\begin{pmatrix} x \\ y \end{pmatrix} = \begin{pmatrix} 12 \\ 21 \end{pmatrix}.$$

$\det \mathbf{A} = 12 - 9 = 3$,

$$\mathbf{A}^{-1} = \frac{1}{3}\begin{pmatrix} 6 & -3 \\ -3 & 2 \end{pmatrix}$$

$$\begin{pmatrix} x \\ y \end{pmatrix} = \frac{1}{3}\begin{pmatrix} 6 & -3 \\ -3 & 2 \end{pmatrix}\begin{pmatrix} 12 \\ 21 \end{pmatrix} = \begin{pmatrix} 3 \\ 2 \end{pmatrix}.$$

Check $2 \times 3 + 3 \times 2 = 12$, $3 \times 3 + 6 \times 2 = 21$.

(ii)
$$\begin{pmatrix} 4 & 5 \\ 12 & 15 \end{pmatrix}\begin{pmatrix} x \\ y \end{pmatrix} = \begin{pmatrix} 17 \\ 18 \end{pmatrix} \quad \text{or} \quad \mathbf{B}\begin{pmatrix} x \\ y \end{pmatrix} = \begin{pmatrix} 17 \\ 18 \end{pmatrix}.$$

$\det \mathbf{B} = 60 - 60 = 0$; there is no unique solution.

The two equations represent parallel lines which cannot intersect; thus no solution exists satisfying both equations simultaneously.

(iii)
$$\begin{pmatrix} 1 & 2 \\ 3 & 6 \end{pmatrix}\begin{pmatrix} x \\ y \end{pmatrix} = \begin{pmatrix} 6 \\ 18 \end{pmatrix} \quad \text{or} \quad \mathbf{C}\begin{pmatrix} x \\ y \end{pmatrix} = \begin{pmatrix} 6 \\ 18 \end{pmatrix}.$$

$\det \mathbf{C} = 6 - 6 = 0$; there is no unique solution. $3x + 6y = 18$ is a multiple of $x + 2y = 6$ and represents the same straight line so the co-ordinates of *every* point on the line satisfy both equations.

SOLUTION BY ELIMINATION AND SUBSTITUTION

These algebraic methods we demonstrate in the example which follows.

Example 5

Solve the equations (i) $2x - 4y = 12$, $3x + 2y = 6$ by elimination, (ii) $x - y = 5$, $2x + 3y = 15$ by substitution.

(i)
$$2x - 4y = 12 \quad (1)$$
$$3x + 2y = 6. \quad (2)$$

Multiply (1) by 3, $6x - 12y = 36$

multiply (2) by 2, $6x + 4y = 12$,

subtract, $-16y = 24$

$$y = -1\tfrac{1}{2};$$

substitute for y in (2), $3x - 3 = 6$

$$3x = 9$$
$$x = 3.$$

Check, by substituting $x = 3$, $y = -1\frac{1}{2}$ in (1); $6 + 6 = 12$.

(ii)
$$x - y = 5 \quad (1)$$
$$2x + 3y = 15. \quad (2)$$

From (1) $x = 5 + y$;

substitute for x in (2)

$$2(5 + y) + 3y = 15$$
$$10 + 2y + 3y = 15$$
$$5y = 5$$
$$y = 1,$$

and $x = 5 + 1 = 6$.

Check, by substituting $x = 6$, $y = 1$ in (2); $12 + 3 = 15$.

NEWTON'S RINGS TO DETERMINE WAVELENGTH, λ

Newton's rings are used in an experiment to determine λ, the wavelength of monochromatic light. These rings are concentric circular interference fringes formed in the air wedge between a lens whose lower surface has radius of curvature R, and a plate of glass.

We use the formula

$$d_n^2 = 4Rn\lambda,$$

where d_n is the diameter of the nth ring.

Usually the centre of the fringe system is blurred, making it difficult to be certain of the exact number n, of a particular fringe, although it is easy to count m fringes away from it. By measuring d_n and d_{n+m}, and knowing m, we obtain two simultaneous equations which we can solve for λ and n. The equations are, for the nth ring,

$$d_n^2 = 4Rn\lambda,$$

and for the $(n + m)$th ring,

$$d_{n+m}^2 = 4R(n + m)\lambda.$$

Exercise B
Where possible, solve the following simultaneous equations.

1. $4x - 2y = 5$
   $6x + 2y = 5$
2. $4x - 2y = 12$
   $6x + 2y = 8$
3. $2x - y = 0$
   $3x + y = 5$
4. $4x + y = 5$
   $8x + 2y = 7$
5. $3x + y = 9$
   $5x + 2y = 16$
6. $4x + 3y = 5$
   $x + y = 2$
7. $y = 2 - 3x$
   $2x = 4y + 5$
8. $x + y = 7$
   $x - y = 5$
9. $x = 3 + 2y$
   $3y = 2x - 10$

Numbers 10–12 concern the results of Newton's rings experiments. In each case solve for n and λ, remembering than n is a whole number. Record the results by copying and completing the table.

	d_n/cm	d_{n+m}/cm	m	R/cm	λ	n
10.	0.40	0.49	3	113		
11.	0.32	0.45	4	109		
12.	0.24	0.43	6	122		

QUADRATIC EQUATIONS

The general quadratic equation is

$$ax^2 + bx + c = 0$$

where a, b, c are constants referred to as the coefficients of x^2 and x, and the independent term (independent of x) respectively. Graphical solution of such equations is possible but tedious and inaccurate; we cannot use flow diagrams directly. But if the expression can be factorised we can use the following argument to find the solutions:

Suppose X and Y are two real numbers and that $X \neq Y$.
Then $XY = 0 \Leftrightarrow$ either $X = 0$ or $Y = 0$.

Example 6
Solve the quadratic equation $x^2 - 5x + 6 = 0$.
Now $x^2 - 5x + 6 = (x - 3)(x - 2)$ so we can write

$$(x - 3)(x - 2) = 0,$$

$\Leftrightarrow$ either $x = 3$ or $x = 2$.

This simple diagram can help with factorising.

Before

	x
x	x^2

Try $6 = -2 \times -3$; after

	x	-2
x	x^2	$-2x$
-3	$-3x$	6

$-2x + (-3x) = -5x$ so this guess gave the correct answer.

Exercise C
Solve the following quadratic equations:

1. $3x^2 - 2x = 0$
2. $x^2 - 3x + 2 = 0$
3. $x^2 + 7x + 10 = 0$
4. $x^2 - 4x - 5 = 0$
5. $x^2 + 3x - 10 = 0$
6. $2x^2 - 9x + 4 = 0$
7. $10/x = 3x + 13$
8. $x^2 = 11x - 28$

GENERAL SOLUTION OF THE QUADRATIC EQUATION

Not all quadratic equations can be solved by simple factorisation, so we need a general method and derive a formula.

$$ax^2 + bx + c = 0;$$

divide by a,
$$x^2 + \frac{bx}{a} + \frac{c}{a} = 0$$

subtract $\frac{c}{a}$ from both sides,
$$x^2 + \frac{bx}{a} = -\frac{c}{a}$$

make the left hand side a perfect square by adding $(b/2a)^2$ to both sides

$$x^2 + \frac{bx}{a} + \left(\frac{b}{2a}\right)^2 = -\frac{c}{a} + \left(\frac{b}{2a}\right)^2$$

$$\left(x + \frac{b}{2a}\right)^2 = \frac{b^2 - 4ac}{4a^2}$$

take the square roots of each side

$$x + \frac{b}{2a} = \frac{\pm\sqrt{b^2 - 4ac}}{2a}$$

subtract $\frac{b}{2a}$ from both sides,

$$x = \frac{-b \pm \sqrt{b^2 - 4ac}}{2a}.$$

This is a very important formula which you should learn; you can use it to solve any quadratic equation, but if $b^2 < 4ac$ the equation has no real roots.

Applications of the general formula for the solution of quadratic equations frequently occur in physics and chemistry. Two applications follow.

CONVERGING LENS

In geometrical optics, it can be shown that there is a minimum distance between the object and its real image in a converging lens. We substitute in the lens formula

$$\frac{1}{u} + \frac{1}{v} = \frac{1}{f}$$

(using the real is positive convention) where u and v are object and image distances respectively and f the focal length. If d is the distance between the object and its image then $d = u + v \Leftrightarrow v = d - u$. Thus

$$\frac{1}{u} + \frac{1}{(d-u)} = \frac{1}{f}$$

$$\Rightarrow \frac{d}{u(d-u)} = \frac{1}{f}$$

$$\Rightarrow u^2 - ud + fd = 0.$$

Using the solution of the quadratic equation for u, then

$$u = \frac{+d \pm \sqrt{d^2 - 4fd}}{2}$$

This equation has real roots provided

$$d^2 \geqslant 4fd$$

$$\text{or} \quad d \geqslant 4f, \quad \text{since} \quad d \neq 0.$$

Thus the minimum distance between the object and its real image in a converging lens is four times the focal length ($d = 4f$).

DETERMINATION OF EQUILIBRIUM CONSTANTS

Consider the equilibrium reaction between ethanoic acid and ethanol forming an ester plus water.

	CH_3COOH +	$C_2H_5OH \rightleftharpoons$	$CH_3COOC_2H_5$ +	H_2O
Initial number of moles	n	m	0	0
Number of moles at equilibrium	$n-x$	$m-x$	x	x

If the total volume of the mixture is V then the concentrations are respectively

$$\frac{n-x}{V} \qquad \frac{m-x}{V} \qquad \frac{x}{V} \qquad \frac{x}{V}$$

The equilibrium constant K is defined as,

$$K = \frac{[CH_3COOC_2H_5][H_2O]}{[CH_3COOH][C_2H_5OH]}$$

$$= \frac{\frac{x}{V} \cdot \frac{x}{V}}{\left(\frac{n-x}{V}\right)\left(\frac{m-x}{V}\right)}$$

$$= \frac{x^2}{(n-x)(m-x)}.$$

Example 7
Given $n = 0.2$ mol, $m = 0.3$ mol and $K = 3.92$ at 298 K. Calculate x, the number of moles of acid (or alcohol) which are converted at equilibrium.

$$3.92 = \frac{x^2}{(0.2 - x)(0.3 - x)}$$

$\Rightarrow 2.92x^2 - 1.96x + 0.235 = 0.$

The solution of the general quadratic equation $ax^2 + bx + c = 0$ is given by

$$x = \frac{-b \pm \sqrt{b^2 - 4ac}}{2a},$$

where $a = 2.92$, $b = -1.96$ and $c = 0.235$ in this example.

By substitution we obtain $x = 0.51$ or 0.16 mol. Since the number of moles of alcohol or acid which are converted cannot exceed the initial amounts provided, the answer required is $x = 0.16$ mol.

Exercise D
Solve the following quadratic equations where possible either by using the formula or by factorising. If the equation has no real roots, say so.

1. $x^2 - 5x + 5 = 0$
2. $x^2 - 3x - 28 = 0$
3. $2x^2 + x - 15 = 0$
4. $x^2 + 3x + 4 = 0$
5. $x^2 + 3x + 2 = 0$
6. $7x^2 = 6x - 5$
7. $6x - \frac{1}{x} - 3 = 0$
8. $3x^2 + 2x - 1 = 0$
9. $x^2 - 3x + 1 = 0$
10. $5x^2 + 2x - 2 = 0$
11. $4x^2 + 12x + 9 = 0$
12. $4x^2 + 12x - 20 = 0$
13. $6t^2 - 20t = 0$
14. $30 - x - x^2 = 0$

Exercise E
1. For the following equilibria calculate the number of moles of each substance present at equilibrium, and derive an expression for the equilibrium constant in a concise form.
(i) Consider $2a$ mol of NH_3 initially, let x be the fraction dissociated.

$$2NH_3(g) \rightleftharpoons N_2(g) + 3H_2(g).$$

(ii) Consider a mol CH_3COOH and b mol C_2H_5OH originally, let x moles of $CH_3COOC_2H_5$ be present at equilibrium.

$$CH_3COOH(l) + C_2H_5OH(l) \rightleftharpoons CH_3COOC_2H_5(l) + H_2O(l).$$

(iii) Given $K = 4$ at 298 K for the above equilibrium, calculate x when $a = b = 3$ moles.

2. A mixture of 2.5×10^{-2} mole of I_2 and 1.19×10^{-2} mole of H_2 was allowed to reach equilibrium at 733 K. Given the equilibrium constant $K = 48.7$, calculate the number of moles of HI present at equilibrium.
3. Calculate the percentage of acid that is esterified when a mixture of 1 mole of ethanoic acid, 1 mole of water and 1 mole of ethanol reaches equilibrium at 373 K. Assume the equilibrium constant $K = 4$ at this temperature.

6

STATISTICS I

STATISTICS, MEAN, MEDIAN AND MODE

A **statistic** is a single number which describes some character of a group of data (a population); the study of such numbers is called Statistics. Understanding of statistics is essential to the proper use of any numerical data collected in experiments, our own or other people's.

We begin by considering three different numbers which can be calculated to describe the central tendency of a population (different definitions of the middle if you like). They are **mean value,** also loosely called the average, or more exactly the arithmetic mean, **median value**, the value of the middle member of the population when they are arranged in order, and **mode**, the most common value. The following example illustrates their use.

Example 1

The number of letters in the words of the first sentence of this chapter are as follows:

1 9 2 1 6 6 5 9 4 9 2 1 5 2 4 1 10 3 5 2 4 7 2 6 10.

Find their mean, median and mode.

Rearrange in ascending order

1 1 1 1 2 2 2 2 2 3 4 4 4 5 5 5 6 6 6 7 9 9 9 10 10.

There are 25 in the population, the 13th number is 4, and the most common is 2.

$$Mean = \frac{4\times1+5\times2+1\times3+3\times4+3\times5+3\times6+1\times7+3\times9+2\times10}{25} \quad \text{(a)}$$

$$= \frac{4+10+3+12+15+18+7+27+20}{25}$$

$$= \frac{116}{25}$$

$$= 4.64.$$

For this population we have

Median = 4 letters, **Mode** = 2 letters, **Mean** = 4.64 letters.

Why three different numbers? They summarise different information, and which to use is a matter of judgement. In this example the *median* is a good descriptive statistic because the mean is rather weighted by a few long words. If the sentence is taken as a sample and used to estimate the number of letters in 1000 words the *mean* must be used. This makes sense of quoting the mean as 4.64 letters (at first sight a foolish statement) because $4.64 \times 1000 = 4640$ letters. However 25 is much too small a sample to justify so accurate an estimate, and, giving the mean to 2 significant figures, 4.6 letters would be more reasonable here. The *mode* is unlikely to be much use in this case.

Fig. 6.1.

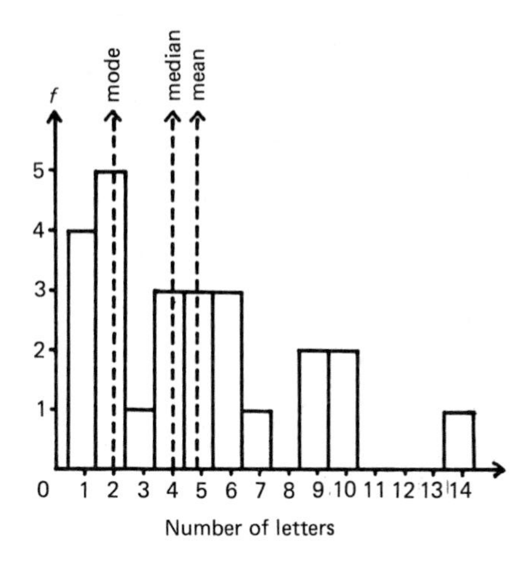

FREQUENCY TABLE AND FREQUENCY DIAGRAM

The data is easier to understand and to process when arranged in a **frequency table**; the data of example 1 is rearranged in this way below.

No. of letters x	*Frequency* f	$f \times x$
1	4	4
2	5	10
3	1	3
4	3	12
5	3	15
6	3	18
7	1	7
8	0	0
9	3	27
10	2	20
	$N = 25$	116

This is a much more convenient arrangement when calculating the mean than (a) above. The total in the $f \times x$ column gives the total number of letters (the numerator in (a)) and the total of the f column is N, the number of words in the sentence (the denominator in (a)).

From this table we can draw a frequency diagram (figure 6.1) to display the data visually. Note that each column is marked at its centre with the number of letters. Its height is determined by the frequency of words of that length, and since all the columns are the same width the area of each is also proportional to the frequency. The mean, mode and median are indicated on the diagram.

GROUPING DATA: CLASS INTERVALS

The distribution discussed in the preceding example was of a discrete variable. There are many such—the number of eggs in a clutch, of bluebell flowers on a stem, of counts on a Geiger counter—when all the members of the population of data are integers. The following example concerns measurements, necessarily given to a certain accuracy, so that each value of the variable x can only be said to lie between certain limits. We then naturally think of x as varying continuously.

Example 2

The heights of the 20 members of the Biology class are listed below, measured in metres to the nearest centimetre. Group the data in 10 cm class-intervals, draw a frequency diagram, calculate the mean and mode.

1.65 1.64 1.59 1.67 1.74 1.59 1.67 1.72 1.59 1.80
1.59 1.52 1.59 1.68 1.64 1.58 1.59 1.67 1.53 1.53

Class-interval/m	Mid-interval mark x	frequency f	fx
1.50 – 1.59	1.545	10	15.450
1.60 – 1.69	1.645	7	11.515
1.70 – 1.79	1.745	2	3.490
1.80 – 1.89	1.845	1	1.845
		20	32.300

The total in the f column is N, here 20.

$$\text{Mean} = \frac{32.300}{20} = 1.615\,\text{m}.$$

Mode = 1.50 – 1.59 m, the most common group, sometimes called the modal group.

The data in this example is fairly widely dispersed and will be easier to comprehend when collected together in 10 cm groups, 1.50–1.59 m, 1.60–1.69 m and so on. Note that every measurement falls into one of the groups, never on to the boundary between two. As you read through the data make a stroke for each in its interval (the technique of drawing the 5th stroke across the preceding 4 is a very good one, widely used, as it reduces errors in the final counting). The mid-interval marks are the mid-points of the intervals, here 1.545 m, 1.645 m and so on (it is often possible to reduce later work by grouping data as you collect it).

You will notice that in the frequency diagram, figure 6.2, the boundaries of the columns are at 1.495 m, 1.595 m and not at 1.50 m, 1.60 m and so on. A height $h = 1.60$ tells you that the class member stands between 1.595 m and 1.605 m high. More precisely, $1.595\,\text{m} \leqslant h < 1.605\,\text{m}$, so that the lower end of the interval whose lowest member is 1.60 m is at 1.595 m.

The calculation of the mean from grouped data makes the assumption that the mean of the members in each class is the mid-interval mark, an approximation which we must realise we are making but which is usually acceptable. For this population the *mean* is a good descriptive statistic, and the *mode* too might be useful (the local stockist would hold more track suits to fit heights 1.50 to 1.59 m). The *median* could be found very quickly by lining up the students in height order, and measuring the height of the middle member of the line. Exercise A subjects the data of example 2 to different treatments to illustrate this. Compare the means you obtain with the value calculated above, 1.615 m, and note the appearance of the frequency diagram when you use 5 cm classes.

Fig. 6.2.

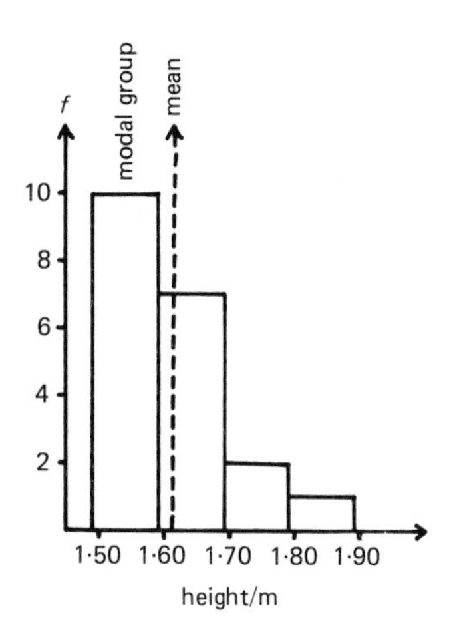

Exercise A

1. Calculate the mean height of the members of the Biology class directly from the original list in example 2.
2. Group the same data using class-intervals 1.50–1.54 m, 1.55–1.59 m and so on. Draw the frequency diagram and calculate the mean and mode.

FREQUENCY DIAGRAMS: SUMMARY

A frequency diagram should show the way in which the population is distributed; too small a class-interval obscures this by showing too much detail, too large a one obscures by smoothing out the detail. The mid-interval marks are the mid-points of the intervals and in frequency diagrams the boundaries must be drawn at the ends of the intervals (for a class-interval of 2.50 to 2.59, for instance, the boundaries are drawn at 2.495 and 2.595).

CUMULATIVE FREQUENCY, MEDIAN, QUARTILES

You may have kept a running total sometime, when scoring at cricket or bridge, or keeping accounts of your savings. This is very similar to compiling a cumulative frequency table which is used to find the median when the population cannot be arranged in order because it is grouped and large. Again we take an example as illustration, but first we need to explain the new term quartile. The median is the middle member of a population, and half the population lies below it. The members of the population at the quarter and three-quarter points are called the **lower quartile** and the **upper quartile**, and you will see that

$\frac{1}{4}$ of the population must have values not more than the lower quartile

$\frac{3}{4}$ of the population must have values not more than the upper quartile.

Example 3

The Biology class whose heights were given in example 2 formed part of a group of 123 students, all female. The heights of the whole group were measured, and the results recorded in the frequency table below (class-intervals 5 cm). Use a cumulative frequency table to plot a cumulative frequency graph and from the latter find the median and the upper and lower quartiles.

Class interval/m	*frequency* f	*cumulative frequency* *c.f.*
1.50–1.54	5	5 not more than 1.545
1.55–1.59	21	26 not more than 1.595
1.60–1.64	37	63 not more than 1.645
1.65–1.69	36	99 not more than 1.695
1.70–1.74	16	115 not more than 1.745
1.75–1.79	7	122 not more than 1.795
1.79–1.84	1	123 not more than 1.845
	$N = 123$	

Figure 6.3 shows both the frequency diagram (a) and the cumulative frequency diagram (b), but there is no need to draw both every time you require a c.f. curve.

To find the median read from the c.f. diagram the height corresponding to c.f. = 61.5. Median = 1.64 m.

$\frac{3}{4} \times 123 = 92.25$. From the graph, upper quartile = 1.68 m.
$\frac{1}{4} \times 123 = 30.75$. From the graph, lower quartile = 1.60 m

Fig. 6.3 (*a*)

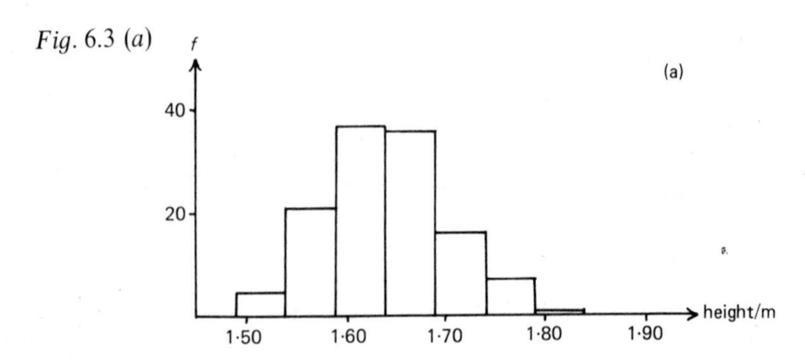

The points on the c.f. curve come at the end of each class interval (as they must, since the total to that point must include all members of the population below it).

Quartiles give a measure of the spread of the distribution of the heights, and 50% of the population lies between 1.68 m and 1.60 m. This idea is easily extended to other percentages, for instance 10% of the population lies above 1.725 m, the height corresponding to

$$\frac{123 \times 9}{10} = 110.7$$

Such statistics are called **percentiles.**

The median bisects the area of the frequency diagram, and 50% of the area lies between the two quartiles.

Fig. 6.3 (*b*)

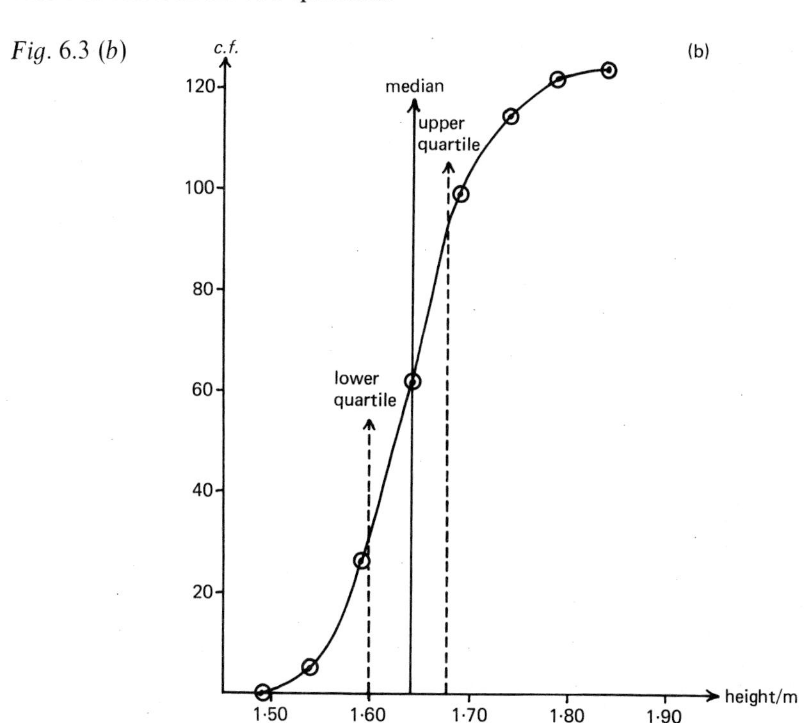

Exercise B

1. In 300 families the number of children were recorded:

No. of children	x	0	1	2	3	4	5
No. of families	f	37	76	93	53	22	19

Find the mean and the mode. Draw the frequency diagram and mark on it the mean.

2. Below are the examination marks for a class. Group them in classes of 10, 1–10, 11–20, and so on. Assign the mid-interval marks, draw the frequency diagram and calculate the mean.
64 66 69 70 63 59 61 50 56 68 50 72 67 71 57 61 76 63 60 54 66.
3. Human height is determined by a large number of genes, and also by certain factors in the environment, notably the diet. In a simplified model of the genetic factors ten coins were tossed and the number of heads recorded; this represented the number of 'genes for tallness', the results being listed below for 50 trials representing 50 individuals.

No. of "genes for tallness"	0	1	2	3	4	5	6	7	9
No. of individuals	0	0	1	3	16	14	10	5	1

Draw a frequency diagram, calculate the mean and the mode of "genes for tallness".
4. Compile a cumulative frequency table for the following examination marks and use it to draw a cumulative frequency diagram. From this graph answer the following questions. What is the median? What are the upper and lower quartiles? 85% of the population lies above a certain value. What is it?

Mark	1–10	11–20	21–30	31–40	41–50	51–60	61–70
f	0	7	14	9	19	12	5

Mark	71–80	81–90	91–100
f	1	2	0

5. Collect your own data. Suggestions are
(a) lengths of pine needles,
(b) number of spines in holly leaves,
(c) pulse rates,
(d) number of florets on composite flower such as dandelion, daisy,
(e) lengths of forefingers.
For each population find the mean and draw the frequency diagram. Keep the data for use in future exercises.

MEASURES OF SPREAD OR DISPERSION

The shape of a distribution shows up well in the frequency diagram but a number is needed to describe its width, or spread. The **range** is the difference between the largest and smallest members of the population, but gives no indication of what happens between these. The **semi-inter-quartile range**, that is (upper quartile—lower quartile)/2, takes more account of the shape of a distribution.

Another approach to the problem is to find the mean, m, and then to average the deviations from it. But this is by definition zero; we overcome this by taking the **positive** value of each deviation, called the **absolute** value and written $|x - m|$, and averaging that. This statistic is called the **mean absolute deviation from the mean.** It is not difficult to calculate and in itself is quite a useful statistic, but for mathematical reasons it is not widely used. We give one example however to show how to calculate it.

Example 4

In 50 throws of two dice the scores were recorded. Calculate the mean and the mean absolute deviation from the mean.

x	f	$f \times x$	$\|x - m\|$	$f \times \|x - m\|$	
2	2	4	4.7	9.4	
3	5	15	3.7	18.5	
4	4	16	2.7	10.8	
5	4	20	1.7	6.8	mean $= m$
6	5	30	0.7	3.5	$= \dfrac{337}{50}$
7	10	70	0.3	3.0	
8	7	56	1.3	9.1	$= 6.7$
9	6	54	2.3	13.8	
10	6	60	3.3	19.8	
11	0	0	4.3	0.0	
12	1	12	5.3	5.3	
	$N = 50$	337		100.0	

$$\text{Mean absolute deviation from the mean} = \frac{100.0}{50} = 2.0.$$

VARIANCE AND STANDARD DEVIATION

By squaring the deviations from the mean we ensure that we have positive numbers to deal with, and it happens that the mean of these squared deviations, called the variance s^2, is very convenient mathematically. To quote a statistic concerning the spread of a distribution of mouse-tail-lengths in cm^2, or of the numbers of eggs in blackbirds' nests in $eggs^2$ makes no sense, so a new statistic is defined. This is the standard deviation $s = +\sqrt{s^2}$, often called S.D. Before investigating s^2 and s further we need to be able to use a new notation.

SIGMA NOTATION

The symbol $\sum$ (sigma) is the Greek capital letter S and means *the sum of.* Thus

(a) $$\sum_{i=1}^{i=4} i = 1 + 2 + 3 + 4 = 10,$$

and

(b) $\sum_{i=1}^{i=5} (i-3)^2 = (1-3)^2 + (2-3)^2 + (3-3)^2 + (4-3)^2 + (5-3)^2$

$$= (-2)^2 + (-1)^2 + 0 + 1^2 + 2^2$$
$$= 4 + 1 + 0 + 1 + 4$$
$$= 10.$$

Thus i takes each integral value between 1 and 4 in (a), between 1 and 5 in (b), including the first and last. Sometimes i is used as a suffix, written x_i. In example 4 the second number is 3, frequency 5, and we write $x_2 = 3$, $f_2 = 5$, reading these as "x-two equals three" and so on. Using this notation we can sum up the calculation in example 4 in a formula, namely

$$\text{Mean absolute deviation from the mean} = \frac{\sum_{i=1}^{i=n} f_i|x_i - m|}{N}$$

(in example 4, n is 11).

Exercise C

1. The frequency table of a distribution is given below; write down x_4, f_2, f_5x_5. Write out in full

$$\sum_{i=1}^{i=6} f_i$$

and calculate it. What have you just computed?

i	1	2	3	4	5	6
x_i	10	15	20	25	30	35
f_i	11	9	11	8	12	10

2. Write out in full and evaluate

(a) $\sum_{i=0}^{i=3} (i+2)$, (b) $\sum_{i=1}^{i=4} i(i-1)$.

3. How many terms are there in the expansion of

$$\sum_{i=5}^{i=16} f_i |x_i - m| ?$$

4. Using the sigma notation write down the formula for the mean of a distribution in two ways: (a) when the frequency table is given as in 1 for n different values, (b) when the observations are listed without any grouping in the order in which they are made and called $x_1, x_2, x_3 \ldots x_n$.

CALCULATION OF VARIANCE AND STANDARD DEVIATION

The **variance** s^2 is the mean of the squares of the deviations from the mean.

The **standard deviation** s or S.D., is the square root of the variance.

In practice this involves adding another column to the table; in example 4 do this for yourself by multiplying each element of the 5th column by the corresponding element of the 4th. You will find that

$$\sum_{i=1}^{11} f_i \times (x_i - m)^2 = 293.70$$

so that

$$s^2 = \frac{293.70}{50} = 5.874, \quad \text{and} \quad s = 2.42.$$

The calculation of variance is simplified if you use another form of the expression. It can be shown that

$$s^2 = \frac{\sum_{i=1}^{n} f_i (x_i - m)^2}{N} = \frac{\sum_{i=1}^{n} f_1 x_i^2}{N} - m^2.$$

We give an example to show how the calculation is laid out, and suggest that you calculate the variance for example 4 using the second form of the formula.

Example 5

Calculate the variance and standard deviation for the population of heights of sixth-formers which is given in example 3.

i	x_i	f_i	$f_i x_i$	$f_i x_i^2$
1	1.52	5	7.6	11.552
2	1.57	21	32.97	51.763
3	1.62	37	59.94	97.103
4	1.67	36	60.12	100.400
5	1.72	16	27.52	47.334
6	1.77	7	12.39	21.930
7	1.82	1	1.83	3.312
		123	202.36	333.394

$$\text{mean} = m = \frac{202.36}{123} = 1.645 \text{ metres}$$

$$m^2 = 2.7060$$

$$\sum_{i=1}^{7} f_i x_i^2 = \frac{333.39}{123} = 2.7105$$

$$s^2 = 2.7105 - 2.7060$$
$$= 0.0045$$
$$s = 0.067 \text{ metres.}$$

In exercise D you are asked to calculate standard deviations for distributions given in exercise B. It is useful to see the spread described by $\pm s$ and by $\pm 2s$ about the mean on the frequency diagrams and so to appreciate the meaning of these ranges. You will find in many cases that most of the members of a population lie within the latter range, about 95% of the members in fact.

SUMMARY

$$\text{Mean} = m = \frac{\sum_{i=1}^{N} x_i}{N} \quad \text{or} \quad m = \frac{\sum_{i=1}^{n} f_i x_i}{N}$$

where N is the total number of the population, $N = \sum_{i=1}^{n} f_i$.

Mode is the most common value.

Median is the middle value.

Cumulative frequency graphs are used to find the median and the percentiles when these are required.

$$\text{Mean absolute deviation from the mean} = \frac{\sum_{i=1}^{n} f_i|x_i - m|}{N} \quad \text{or} \quad \frac{\sum_{i=1}^{N} |x_i - m|}{N}$$

$$s^2 = \text{Variance} = \frac{\sum_{i=1}^{n} f_i x_i^2}{N} - m^2,$$

or

$$s^2 = \frac{\sum_{i=1}^{n} f_i (x_i - m)^2}{N}$$

Standard deviation, $s = \sqrt{\text{variance}}$.

Exercise D

1–5. Find the variance and standard deviation for all the examples in exercise B and on the frequency diagrams for these mark the $\pm s$ and $\pm 2s$ points about the mean.

6. A microscope slide was covered with a square grid and the number of dust particles in each square recorded. Calculate the mean, variance and standard deviation. Draw the frequency diagram.

No. of dust particles	0	1	2	3	4	5
f	38	75	89	54	20	19

7. The number of goals scored by football teams in a league are recorded below. Draw a frequency diagram and calculate mean and standard deviation.

No. of goals	0	1	2	3	4	5	6	7
No. of matches	95	151	115	71	31	124	14	1

HISTOGRAMS

If we take care to draw frequency diagrams in which the area of each block is proportional to the frequency it represents they can be called **histograms.** The diagrams in this chapter have all had equal class intervals, but data is not always so tabulated. Example 6 illustrates this.

Example 6

Plot a histogram showing the following data; draw also (for comparison only) the frequency diagram.

No. of eggs in clutch	1	2	3	4	5 or 6
Frequency, f	3	7	9	6	5

Figure 6.4(a) shows the frequency diagram; while the areas of the columns

Fig. 6.4 (*a*)

Fig. 6.4 (*b*)

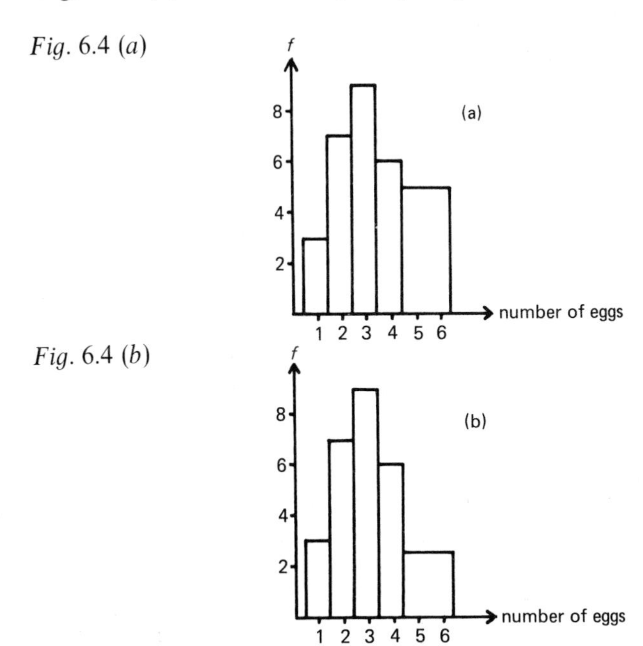

1, 2, 3, 4 are 3, 7, 9, 6 sq. units respectively, note that the area of the final column is 10 sq. units and not 5. In figures 6.4(b) however the area of the

final column is 5 sq. units and this is the histogram we wanted. The scale on the y axis is now frequency density, which is defined by

$$\text{frequenty density} = \text{frequency}/(\text{class-interval}).$$

In this case the class-interval is 1 for 1, 2, 3, 4, and 2 for the final group so that we have the table

No. of eggs	1	2	3	4	5 or 6
Frequency density	3	7	9	6	2.5

Thus in a histogram frequency-density is plotted against the classes, and the area in any interval is proportional to the frequency with which that value occurs.

7

TRIGONOMETRY

Trigonometry, if we consider only the derivation of the word, concerns the measurement of triangles, and this is one aspect of the subject. However, it now covers much more than this, and attention focuses on certain ratios (sine for instance) peculiar to angles. It is therefore appropriate to begin this chapter by discussing the units in which angles are measured.

UNITS OF ANGLES

The first point to note is that angles, and therefore their units, have no dimensions. Length, mass and time are the basic dimensions and appear in compound units, such as ms^{-2}, the unit for acceleration, but an angle involves the ratio of two lengths and hence is dimensionless.

DEGREES

In the full circle there are 360 degrees, written 360° and each degree is subdivided into 60 minutes, written 60′, each minute into 60 seconds, 60″ (strange to think that this unit is inherited from the Babylonians, but so it is). Angles may be expressed as decimals, for example not 27°45′ but 27.75°. Some tables are arranged in this way and most calculators which have trigonometrical facilities require angles to be entered thus.

Fig. 7.1.

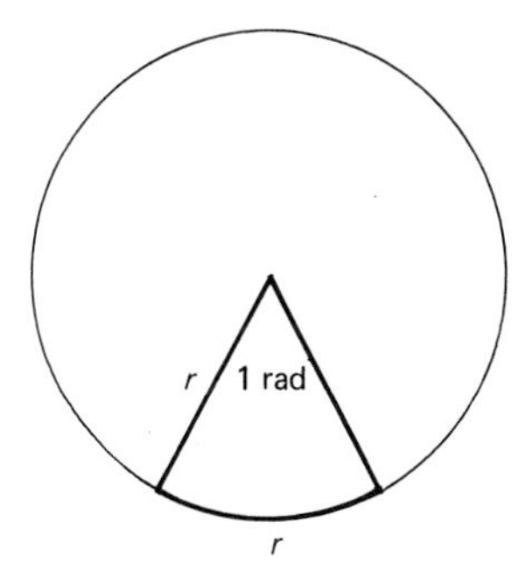

RADIANS

There is another unit of angle which is defined in such a way that it is directly related to the circle; **one radian** is the angle subtended at the centre of a circle by an arc whose length is equal to the radius of the circle (see figure 7.1). Thus

$$360° = 2\pi \text{ radians},$$

$$90^\circ = \frac{\pi}{2} \text{ radians, and in general}$$

$$\theta^\circ = \frac{\pi\theta}{180} \text{ radians.}$$

$$1 \text{ radian} \approx 57.3^\circ.$$

When using calculus angles are always expressed in radians whether the fact is mentioned or not. It is often necessary to change between degrees and radians, so some examples for practice are given in exercise A.

Exercise A

1. Express the following angles in radians to 2 decimal places:
   43°, 21°, 70°, 197°, 243°, 323°.
2. Express the following angles (all given in radians) in degrees, to 2 decimal places:
   0.1, 0.5, 0.67, 1.4, 2.6, 3.4, radians.
3. Express in radians, as multiples of π, the following angles:
   30°, 60°, 120°, 150°, 210°, 240°, 300°, 330°,
   45°, 135°, 225°, 315°,
   90°, 180°, 270°, 360°.
4. Express the following angles, expressed in radians, in degrees:
   $\frac{\pi}{8}, \frac{5\pi}{16}, \frac{3\pi}{20}, \frac{7\pi}{8}$, 0.3, 2, 5.62.

SMALL ANGLES

Figure 7.2 shows the graphs of $y = \theta$, $y = \sin\theta$ and $y = \tan\theta$ for small values of θ, the unit of angle being the radian. The gradient of $y = \sin\theta$ is 1 when $\theta = 0$ which means that for small values of θ, $\sin\theta/\theta \approx 1$; for small angles expressed in radians we can use the approximation $\sin\theta \approx \theta$. In deriving the equations of motion for a simple pendulum, we can use this approximation for angles of swing up to 10° as the following table shows. In it corresponding values of the angle θ expressed in degrees and in radians are given together with $\sin\theta$.

$\theta/^\circ$	0.0000	1.7189	3.4377	4.0107	5.7296	11.4592
θ/radians	0.0000	0.0300	0.0600	0.0700	0.1000	0.2000
$\sin\theta$	0.0000	0.0300	0.0600	0.0699	0.0998	0.1987

$$\text{At } \theta = 0.1 \text{ radians, } \sin\theta - \theta = 0.1000 - 0.0998 = 0.0002,$$

an error of 0.2%, and even at $\theta = 0.2$ radians (approximately 11.5°) $\sin\theta = 0.1987$ the difference is only 0.0013, percentage error 0.67%.

It is also true that for angles expressed in radians the gradient of $y = \tan\theta$ is 1 at $\theta = 0$ (see figure 7.2). For small angles we have another approximation, namely

$$\tan\theta \approx \theta$$

Fig. 7.2.

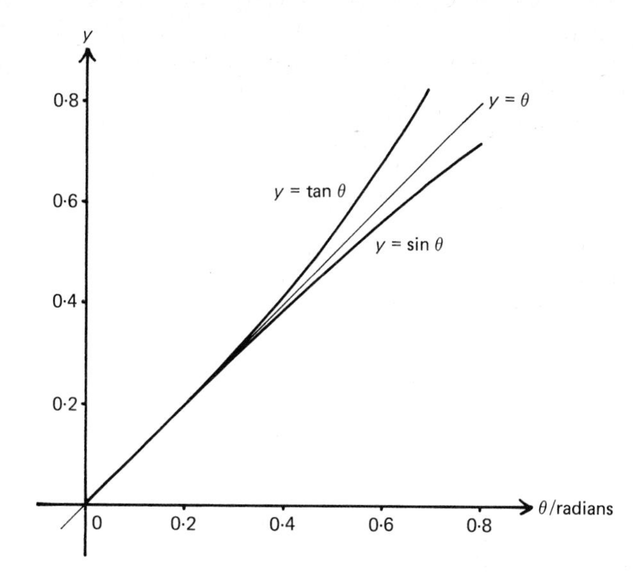

Fig. 7.3.

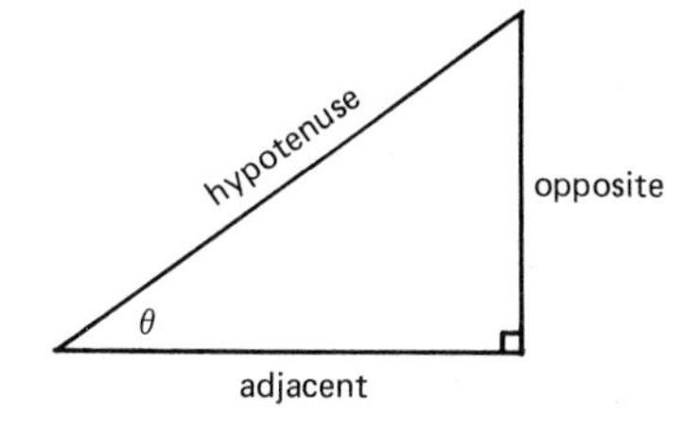

TRIGONOMETRICAL RATIOS OF ANY ANGLE

The definitions $\sin\theta = \dfrac{\text{opp}}{\text{hyp}}$, $\cos\theta = \dfrac{\text{adj}}{\text{hyp}}$, $\tan\theta = \dfrac{\text{opp}}{\text{adj}}$

(see figure 7.3) are helpful mnemonics until the angle becomes greater than one right angle ($\pi/2$ radians), when they become meaningless. They are special cases of the definitions given below, to be read in conjunction with figure 7.4.

$$\sin\theta = \frac{\text{sideways displacement}}{r}$$

$$\cos\theta = \frac{\text{central displacement}}{r}$$

$$\tan\theta = \frac{\sin\theta}{\cos\theta} = \frac{\text{sideways displacement}}{\text{central displacement}}.$$

The angle is always measured in an anti-clockwise direction from the positive x axis OX, and r is always positive; otherwise positive and negative are in the senses usual in Cartesian graphs.

You are aware that tables give the ratios only for $0 \leqslant \theta \leqslant 90°$; figure 7.5 shows quickly, and in a form easily memorised, when they are positive or negative. Figure 7.6 should help you to look up the correct angle in the tables. Returning to figure 7.4, it is obvious that after the radius vector r has turned through an angle of 2π the pattern repeats itself. Expressed in mathematical terms, for k any integer,

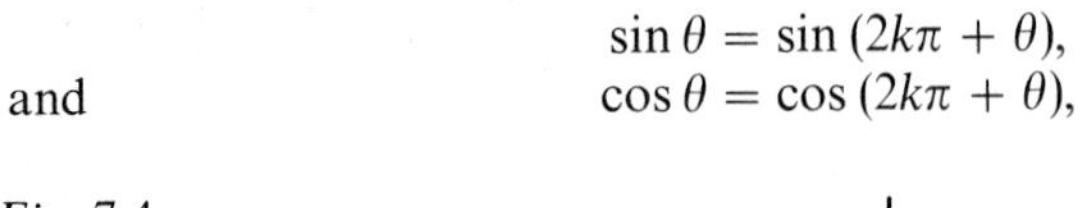

$$\sin\theta = \sin(2k\pi + \theta),$$

and

$$\cos\theta = \cos(2k\pi + \theta),$$

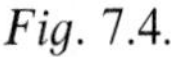

Fig. 7.4.

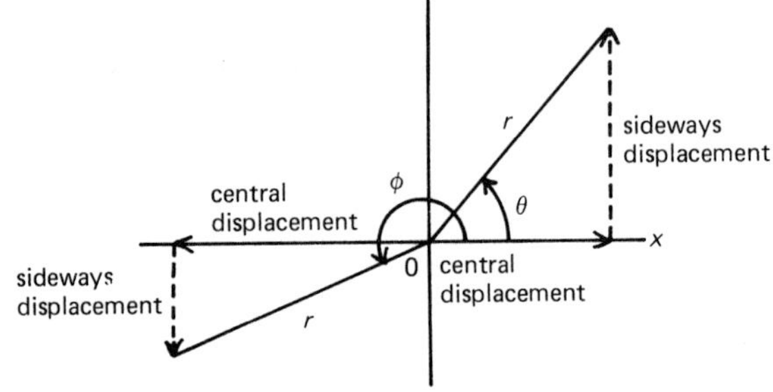

Fig. 7.5.

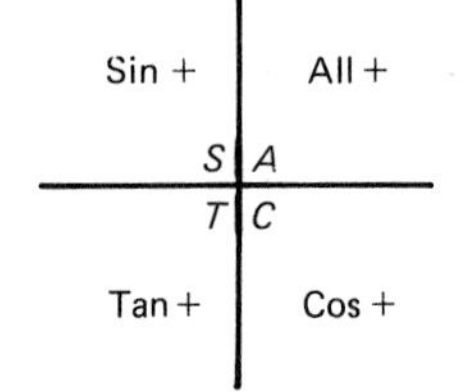

Fig. 7.6.

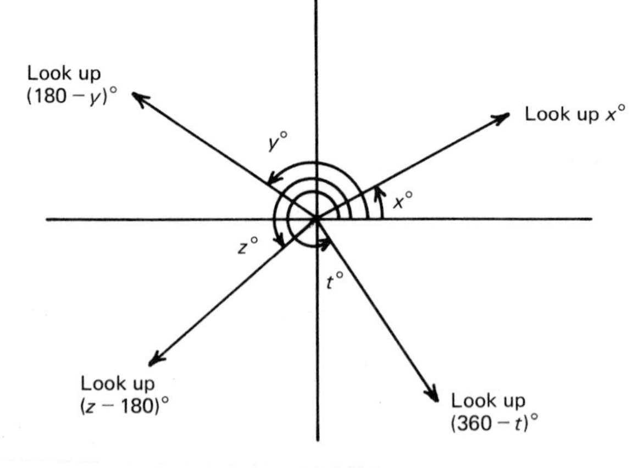

and we say that the trigonometrical functions are **periodic**. Figure 7.7 shows the graphs of $y = \sin x$, $y = \cos x$ and $y = \tan x$. You will see that for the first two, one complete cycle is completed in 2π radians, while for $y = \tan x$ the length of the cycle is π radians. Many different symbols are commonly used for the angle such as θ, x, A, β, α...

NOTES ON TRIGONOMETRIC FUNCTIONS

1. $\cos x$ is an **even function**; that is for all angles, $\cos(-x) = \cos x$, and the graph of $y = \cos x$ is symmetrical about $x = 0$.

2. $\sin x$ is an **odd function**; that is for all angles, $\sin(-x) = -\sin x$, and the graph is not symmetrical about $x = 0$. However, a half-turn about the origin will leave it unchanged.

3. The graphs of $y = \sin x$ and $y = \cos x$ are identical in shape, but displaced by $\pi/2$ in the x direction. In other words, a translation described

Fig. 7.7.

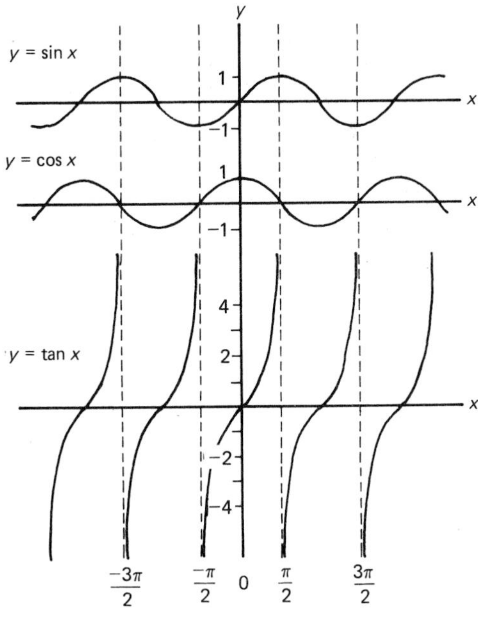

by the column vector $\begin{pmatrix} \pi/2 \\ 0 \end{pmatrix}$ or $\begin{pmatrix} -3\pi/2 \\ 0 \end{pmatrix}$

maps $y = \cos x$ on to $y = \sin x$.

4. You can check from the graphs that

$$\sin\left(\frac{\pi}{2} - \theta\right) = \cos\theta = \sin\left(\frac{\pi}{2} + \theta\right).$$

5. For any real value of θ, $-1 \leqslant \sin\theta \leqslant 1$, and $-1 \leqslant \cos\theta \leqslant 1$. For the domain of all the real numbers the range is the real numbers between -1 and 1.

6. $\tan\theta$ has no such limitations; for the domain $-\frac{\pi}{2} < \theta < \frac{\pi}{2}$, $\tan\theta$ has as range all the real numbers. As the pattern repeats every π radians, the same is true for

$$-\frac{\pi}{2} + k\pi < \theta < \frac{\pi}{2} + k\pi.$$ (k an integer) and we can say that

$$\tan(\theta + k\pi) = \tan\theta.$$

7. For all values of θ, $\cos^2\theta + \sin^2\theta = 1$.

INVERSE TRIGONOMETRIC FUNCTIONS

Given a value of the sine, cosine or tangent there is an infinite number of angles corresponding to it. However in practice it is usually clear which solution is needed and these inverse functions are much used.

Example 1

In a right-angled triangle we know that $\cos A = 0.5$. Find A.

Here $$\cos A = 0.5, \quad A = \frac{\pi}{3} \text{ (or } 60^\circ\text{)},$$

and this is the only solution.

Fig. 7.8.

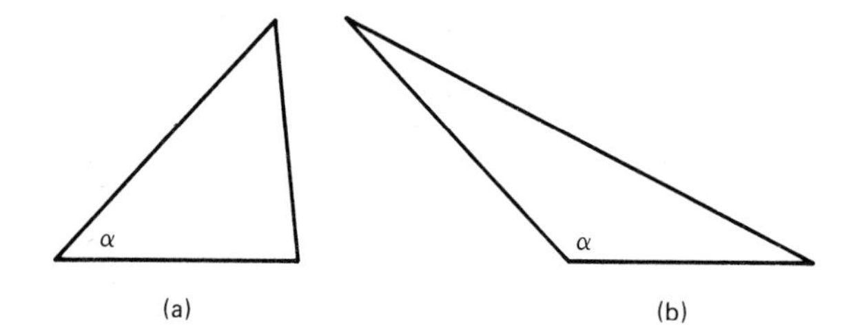

Example 2

In a triangle which is not right angled $\sin \alpha = 0.73$. Find α.

$$\sin \alpha = 0.73$$
$$\Rightarrow \alpha = 46.88^\circ\ (46^\circ 53') \text{ or } 133.12^\circ\ (133^\circ 7').$$

This we can say because $0 < \alpha < 180^\circ$, but we do not know which result to choose. There are in fact two solutions corresponding to the two triangles shown in figure 7.8.

There are two ways of writing the inverse trigonometrical relations; in example 2 for instance, we can write either $\alpha = \sin^{-1} 0.73$ (sine inverse 0.73), or $\alpha = \arcsin 0.73$, read as arc sine 0.73.

Both are often used; some scientific calculators have a key labelling ARC, others INV. Obviously you consult the instructions for your model but generally the answer which appears on the display will be the one in the range

$$-\pi/2 \leqslant \alpha \leqslant +\pi/2, \quad -90^\circ \leqslant \alpha \leqslant +90^\circ \text{ for sine}$$
$$0 \leqslant \alpha \leqslant \pi, \quad 0^\circ \leqslant \alpha \leqslant +180^\circ \text{ for cosine}$$
$$-\pi/2 < \alpha < \pi/2, \quad 90^\circ < \alpha < +90^\circ \text{ for tangent.}$$

Check whether you enter angles in radians or degrees (the latter more likely). The graphs of $y = \sin^{-1} x$, $y = \cos^{-1} x$ and $y = \tan^{-1} x$ over these ranges are shown in figure 7.9.

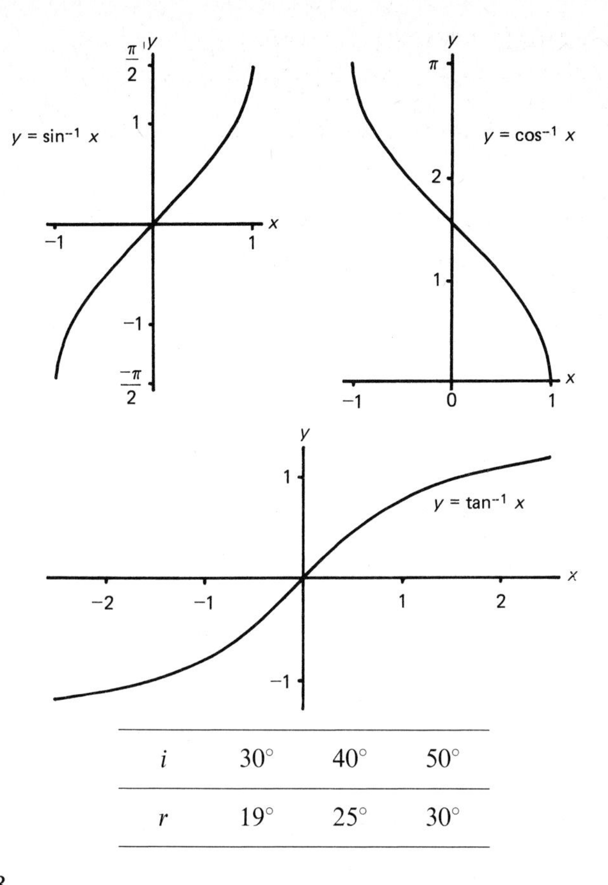

Fig. 7.9.

i	30°	40°	50°
r	19°	25°	30°

Exercise B

1. In an experiment to find the refractive index, n, of perspex, the incident angle, i, in air and the refracted angle, r, in the perspex were measured and tabulated as follows. Use these results to compute values of the refractive index, n, which is given by the expression

$$n = \frac{\sin i}{\sin r}.$$

Then find an average value for n.

2. For total internal reflection the critical angle, c, for a medium is given by

$$n = \frac{1}{\sin c},$$

where n is the refractive index of the medium.

Calculate the critical angle for the media listed below.

medium	Diamond	Crown glass	Flint glass	Olive oil
refractive index	2.42	1.52	1.65	1.46

USEFUL INFORMATION AND FORMULAE

ANGLES OF 30°, 60° AND 45°

The two triangles shown in figure 7.10 give the ratios for these angles directly. (a) is an equilateral triangle with side 2 units, (b) an isosceles right-angled triangle. In many problems it is sufficient to leave the ratios in the form given below.

$$\sin 45^\circ = \frac{1}{\sqrt{2}}; \qquad \sin 30^\circ = \frac{1}{2}; \qquad \sin 60^\circ = \frac{\sqrt{3}}{2}$$

$$\cos 45^\circ = \frac{1}{\sqrt{2}}; \qquad \cos 30^\circ = \frac{\sqrt{3}}{2}; \qquad \cos 60^\circ = \frac{1}{2}$$

$$\tan 45^\circ = 1; \qquad \tan 30^\circ = \frac{1}{\sqrt{3}}; \qquad \tan 60^\circ = \sqrt{3}$$

Fig. 7.10

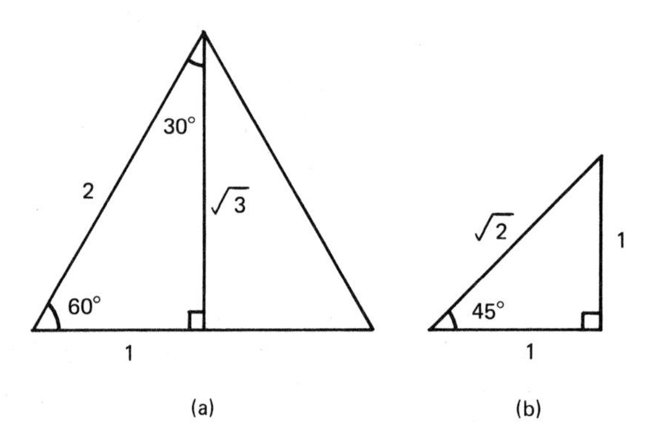

COSECANT, SECANT, COTANGENT

These ratios are defined as

$$\begin{aligned} \operatorname{cosec} \theta &= 1/\sin \theta, \\ \sec \theta &= 1/\cos \theta, \\ \cot \theta &= 1/\tan \theta. \end{aligned}$$

It is as well to recognize them although you may not use them frequently.

SOLUTION OF TRIANGLES

In a triangle labelled as in figure 7.11 the following formulae hold and enable unknown sides and angles to be calculated.

Sine formula $$\frac{\sin A}{a} = \frac{\sin B}{b} = \frac{\sin C}{c}.$$

Fig. 7.11.

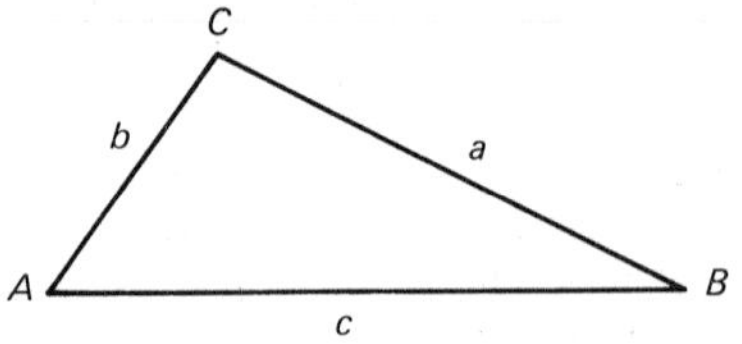

C
b
a
A
B
c

Cosine formula $a^2 = b^2 + c^2 - 2bc \cos C.$

Area of triangle $= \frac{1}{2}bc \sin A.$

Choose the formula which best suits the data, and which involves the least computation.

COMPOUND ANGLES

Given two angles A and B it is a simple matter with tables or a calculator to find $\sin(A + B)$ or $\cos(A + B)$; you *first* add A and B then find the sine or cosine of the sum (first is emphasised because a few trials will demonstrate that finding the sines first and adding them does not work). In some theoretical work however you need the formulae for compound angles which we give below, without deriving them:

$$\sin(A + B) = \sin A \cos B + \cos A \sin B$$
$$\sin(A - B) = \sin A \cos B - \cos A \sin B$$
$$\cos(A + B) = \cos A \cos B - \sin A \sin B$$
$$\cos(A - B) = \cos A \cos B + \sin A \sin B.$$

A useful expression derived from these is

$$\sin A + \sin B = 2 \sin \tfrac{1}{2}(A + B) \cos \tfrac{1}{2}(A - B).$$

VECTORS

Many of you will be familiar with vectors and their notation, but for those who are not a few basic facts are given here. A **vector** has direction as well as magnitude. It is represented by a line drawn to scale to give magnitude, with an arrow to give the direction of the vector; in contrast to this a **scalar** quantity has magnitude only. Force, velocity and momentum are vector quantities; mass and energy are scalar quantities. In printed matter vectors are shown in heavy type, **b** (magnitude b), or sometimes $\overrightarrow{AB}$ (magnitude AB); when writing they are underlined using a wavy line.

Fig. 7.12.

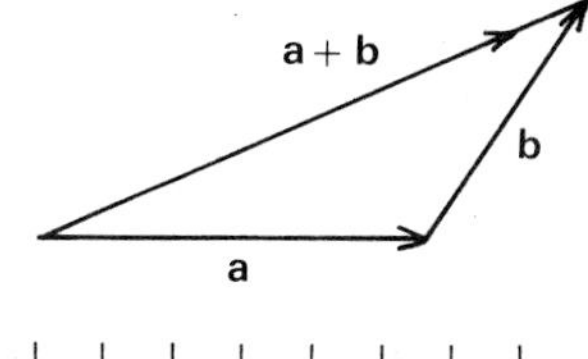

Fig. 7.13.

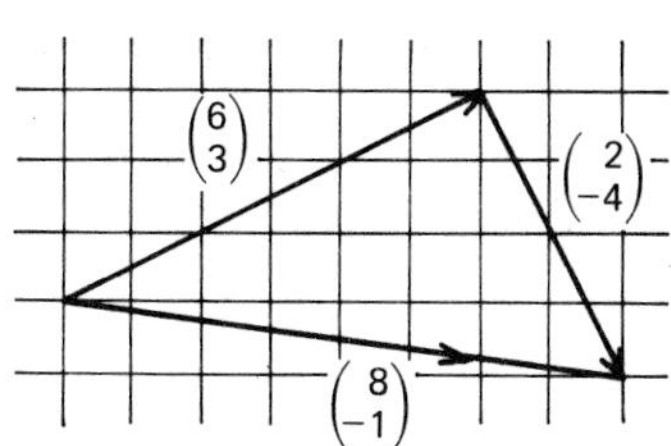

VECTOR ADDITION

Vector addition is shown in figure 7.12 where **a + b** and **a, b** are clearly marked. The well-known Parallelogram Law for the composition of forces is an example of vector addition.

Figure 7.13 shows how to represent vectors in another way using a **column vector**. Using the same convention as in Cartesian graphs, with the same positive directions, we write the vector **p** as $\begin{pmatrix} x \\ y \end{pmatrix}$.

A translation is described quickly by a vector in this way and addition can be done without drawing. In figure 7.13 for instance

$$\mathbf{a} + \mathbf{b} = \begin{pmatrix} 6 \\ 3 \end{pmatrix} + \begin{pmatrix} 2 \\ -4 \end{pmatrix} = \begin{pmatrix} 8 \\ -1 \end{pmatrix}.$$

Magnitude of $\mathbf{a} = a = \sqrt{6^2 + 3^2} = \sqrt{36 + 9} = \sqrt{45} = 6.71$ units.

Fig. 7.14.

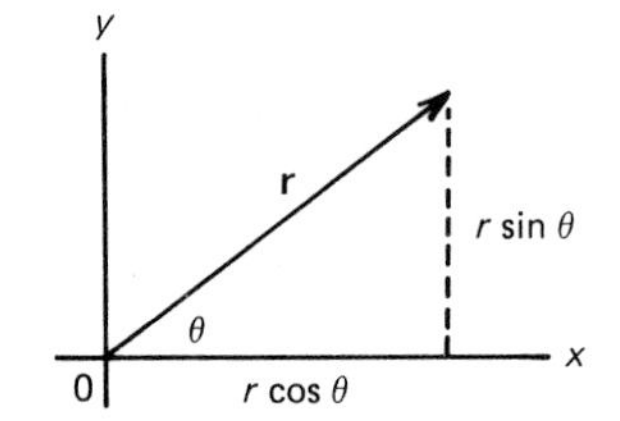

Fig. 7.15.

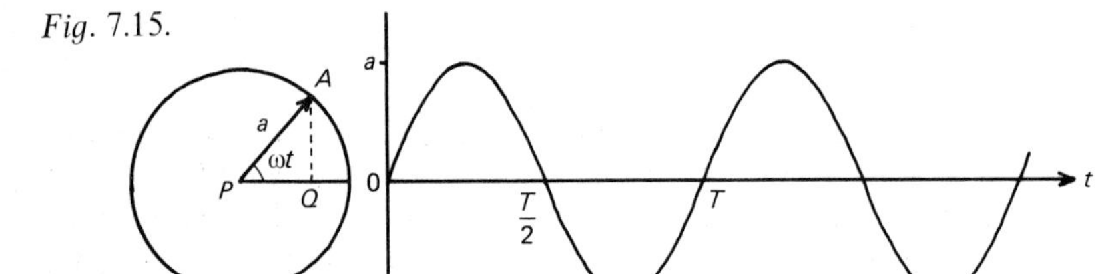

RESOLUTION OF VECTORS

Suppose that **r** is a vector, magnitude r, and that we need to know its components in the directions $0x$ and $0y$ (see figure 7.14). We see that

$$\mathbf{r} = \begin{pmatrix} r\cos\theta \\ r\sin\theta \end{pmatrix}.$$

$r\cos\theta$ is called the component of **r** in the direction $0x$, $r\sin\theta$ the component of **r** in the direction $0y$.

PHASORS OR ROTATING VECTORS

Phasors can be used to represent a sinusoidally varying quantity. Figure 7.15 shows the graph of $y = a\sin\omega t$ where a is the maximum value (amplitude) attained by y.

ω is the **angular frequency** ($\omega = 2\pi f$, where f is the frequency). Alongside the graph is the phasor or rotating vector **PA** which can be thought of as generating the sinusoidal function. **PA** has length (magnitude), a and rotates in an anticlockwise direction with uniform angular velocity ω. At time $t = 0$, **PA** lies along PO; after time t it has rotated through an angle ωt and is in position PA. Then

$$\mathbf{PA} = \begin{pmatrix} a\cos\omega t \\ a\cos\omega t \end{pmatrix} = \begin{pmatrix} PQ \\ QA \end{pmatrix}.$$

Thus QA, the y-component of **PA**, gives the value of the sinusoidal quantity y at any time t. This approach is particularly useful in considering simple harmonic motion (S.H.M.) and alternating current problems.

ANGULAR FREQUENCY AND PERIOD

Further consideration of figure 7.15 shows that **PA** returns to its original position when it has turned through 2π radians.

If this takes time T we have

$$\omega T = 2\pi$$
$$\Rightarrow T = 2\pi/\omega$$

and T is called the **period** of the motion, or the **periodic time**.

Substituting $\omega = 2\pi f$ we have

$$2\pi f T = 2\pi,$$
$$\Rightarrow T = 1/f.$$

SINUSOIDAL VIBRATIONS; PHASE

When a vibration described by $y = a\sin\omega t$ spreads into a medium as a wave the pattern repeats itself periodically in space in a distance λ, called the wavelength. Figure 7.16 shows the wave profile.

At $x = 0$, the oscillation is given by the equation $y = a\sin\omega t$ but the oscillation at A is out of step with the oscillation at the origin. It is said to be out of phase and this must be taken into account when the oscillation at A is described mathematically.

Fig. 7.16.

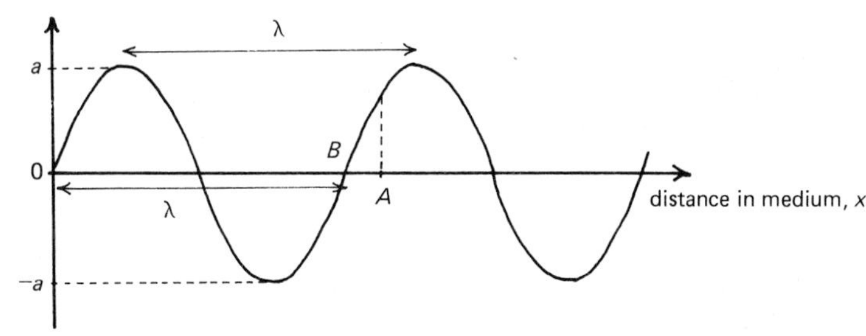

At B, distance λ away from the origin, the angle has changed by 2π radians. Therefore at distance, x, from the origin the angle has changed by $2\pi x/\lambda = \beta x$, where $\beta = 2\pi/\lambda$. βx is called the **phase angle** and the plane progressive wave equation is written as

$$y = a\sin(\omega t - \beta x);$$

the negative sign is used since the vibration at A lags behind that at the origin, O.

It can be shown that $\omega/\beta = v$, where v is the speed with which the wave profile travels through the medium.

Exercise C

1. The wavelength of a progressive wave is 5×10^{-8}m. Calculate the phase difference between two points distance apart (a) 3.06×10^{-7} m, (b) 3.06×10^{-10} m.
2. The e.m.f., E, of a simple dynamo varies with time, t, according to the expression $E = E_0 \sin(\omega t)$, where E_0 is the amplitude or maximum value of the induced e.m.f. and ω is the angular frequency of rotation of the coil.

   For a particular dynamo, $E_0 = 0.5$ volts and $\omega = 100\pi$ rad s^{-1}.

   Calculate the size of E for time, $t = 0.0025$ s, 0.005 s, 0.0075 s, 0.01 s, 0.015 s, 0.02 s.
3. Resolve the following forces vertically and horizontally:
   (a) 20 N acting at 30° to the horizontal.
   (b) 0.05 N acting at 75° to the vertical.
   (c) 220 N along the vertical.

8

CALCULUS, DIFFERENTIATION

RATES OF CHANGE

A car driver varies his speed (rate of change of position measured in km h^{-1}) and his acceleration (rate of change of speed km h^{-2}), adjusting both from moment to moment. He might say "I touched 70 just now", or "I put my foot hard down to pass", tacitly accepting the idea of rate of change at an instant. At the end of the journey he may say that he averaged $65\,\text{km h}^{-1}$ which again is a familiar concept, being (total distance)/(total time). In this chapter we set out first to give a strictly defined meaning to rate of change at a point (or instant), and then to develop the mathematical technique to deal with simple problems.

It would be possible to record the speed of the car at stated intervals, to plot a graph of speed against time, and from it to find the rate of change of speed by drawing the tangent to the curve at the time when we require it. However, it will be more useful to study a curve whose equation is a simple one and this we do in example 1.

Example 1

Draw a graph of $y = x^2$ from $x = -3$ to $x = 3$. Find by calculation the gradients of the chords joining

$P\,(2,4)$ to (a) $P_1\,(3,9)$, (b) $P_2\,(2.5, 6.25)$, (c) $P_3\,(2.2, 4.84)$, (d) $P_4\,(2.1, 4.41)$, (e) $P_5\,(2.01, 4.0401)$.

On the graph draw PP_1, PP_2, PP_4. Draw the tangent to the graph at P and find its gradient.

The graph is shown in figure 8.1.

DRAWING A TANGENT

You will find it useful to have a transparent ruler for drawing the tangent. If the curve is fairly symmetrical about the point of contact (as it is about P in this instance) choose 2 points equidistant from P on the curve and set the ruler across them. Then move it parallel to itself until it just touches the curve, when you can draw a good tangent. When the curve is asymmetrical about the point of contact it helps to oscillate the ruler gently about P until the angles between ruler and curve on each side of P are roughly equal. The two situations are shown in figures 8.2 (a) and (b).

Fig. 8.1.

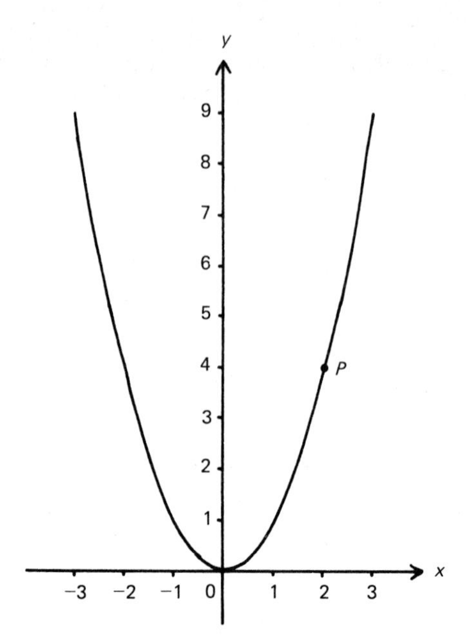

Fig. 8.2.

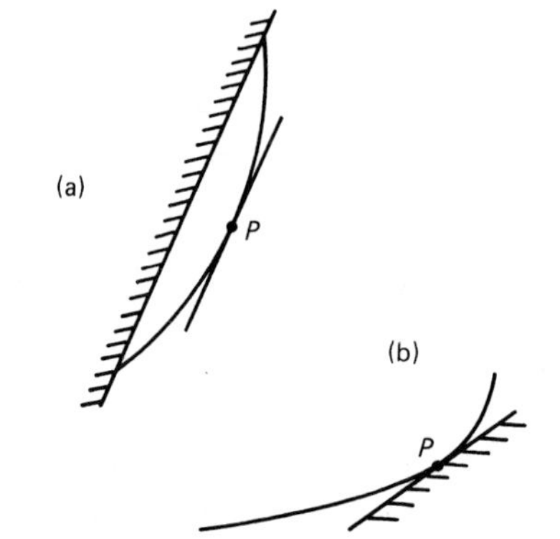

Chord	*Gradient (calculated)*				
PP_1	(9 – 4)/(3 – 2)	=	5/1	=	5.00
PP_2	(6.25 – 4)/(2.5 – 2)	=	2.25/0.5	=	4.50
PP_3	(4.84 – 4)/(2.2 – 2)	=	0.84/0.2	=	4.20
PP_4	(4.41 – 4)/(2.1 – 2)	=	0.41/0.1	=	4.10
PP_5	(4.0401 – 4)/(2.01 – 2)	=	0.0401/0.01	=	4.01

The right-hand column shows clearly that as the ends of the chords approach each other the gradient of the chord tends to 4. This is the value we would expect to find for the gradient of the tangent at *P*. However, drawing that tangent requires skill and you may not find the value from your drawing to be 4.

If you take the point (1, 1) on $y = x^2$ and join it to a series of points on the curve whose x co-ordinates are 2, 1.5, 1.2, 1.1, 1.01, the gradients will be found to approach 2; joining (3,9) to a similarly spaced series of points on the curve will give gradients tending to 6 as the points approach (3,9). This suggests that the gradient of the tangent at (x, x^2) is $2x$.

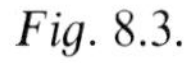
Fig. 8.3.

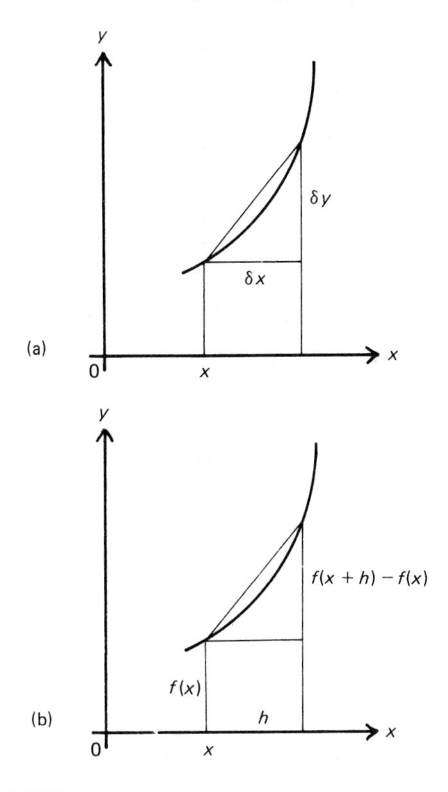

LIMIT, DERIVATIVE

There is an important concept here, that of a **limit,** and using it we can restate our working theory, saying "the limit of the gradient of the chord joining $P(x, x^2)$ to $Q(x+h, (x+h)^2)$, as h tends to zero, is $2x$. This is the **gradient** of the curve $y = x^2$ at P. This sentence is expressed mathematically as follows: when

$$f(x) = x^2$$

$$\lim_{h \to 0} \left\{ \frac{f(x+h) - f(x)}{h} \right\} = \lim_{h \to 0} (2x + h)$$

$$= 2x. \qquad (1)$$

Or

$$\lim_{\delta x \to 0} \frac{\delta y}{\delta x} = 2x \qquad (2)$$

where δy and δx, delta y and delta x, denote the increments in y and x respectively, see figure 8.3(a). (The capital form delta appeared in the work on errors in chapter 2, Δx and Δy.) The increment in x is also commonly called h, so that $\delta x = h$.

Using this notation the increment in y is

$$\delta y = f(x+h) - f(x),$$

see figure 8.3(b). The limit itself is written

$$\frac{\mathrm{d}y}{\mathrm{d}x},$$

always read as "dee y by dee x", and is called the derivative of y with respect to x.

Equation (2) becomes, when $y = x^2$:

$$\frac{\mathrm{d}y}{\mathrm{d}x} = \lim_{\delta x \to 0} \frac{\delta y}{\delta x} = 2x.$$

$f'(x)$ is also commonly used to denote the derivative of $f(x)$.

We must now justify this theory mathematically; let $y = f(x) = x^2$, and suppose P is the point (x, y) on the curve, so that P has co-ordinates (x, x^2). Increase x by h, then $y = (x+h)^2$, and P_1 is the point $(x+h, (x+h)^2)$. We call the increment in x, δx, and the increment in y, δy.

$$\text{Then gradient of } PP_1 = \frac{\delta y}{\delta x} = \frac{(x+h)^2 - x^2}{h}$$

$$= \frac{x^2 + 2xh + h^2 - x^2}{h}$$

$$= \frac{2xh + h^2}{h}$$

$$= 2x + h.$$

But the gradient of the tangent to the curve is the gradient of PP_1 when P_1 approaches P, that is when $h \to 0$ (or $\delta x \to 0$). We have, then,

$$\text{the gradient of the tangent at } P = \lim_{\delta x \to 0} \frac{\delta y}{\delta x}$$

$$\lim_{\delta x \to 0} \frac{\delta y}{\delta x} = \frac{\mathrm{d}y}{\mathrm{d}x} = \lim_{h \to 0} (2x + h) = 2x.$$

In the general case, where $y = f(x)$, we can now state that the derivative of y with respect to x is given by the equation

$$\frac{\mathrm{d}y}{\mathrm{d}x} = \lim_{\delta x \to 0} \frac{\delta y}{\delta x} = \lim_{h \to 0} \frac{f(x+h) - f(x)}{h}.$$

This is shown graphically in figure 8.3(a) and (b). The operation of taking the derivative of y with respect to x is called **differentiation**.

Exercise A

1. Using the data in the following table, plot a graph of temperature against time for a body cooling in surroundings at 24°C. Use the graph to find the time-rate of fall of temperature (the gradient) at several values of temperature. Then use these rates to plot a graph of rate of cooling against temperature above the surroundings.

What conclusions can you draw from this last graph?

Time/min	0	1	2	3	4	5	6	7	8
Temperature/°C	131.0	123.0	115.5	108.5	102.0	96.0	90.5	85.5	81.0

Time/min	9	10	11	12	13	14	15
Temperature/°C	76.0	71.0	68.5	65.0	62.0	59.0	56.0

2. For $y = x^2$ calculate $f(1 + 0.1)$, $f(2 + 0.5)$, $f(2 + 0.001)$, $f(3)$, $f(3 + 0.2)$.
3. Use the results of (2) to calculate the gradients of the chords joining the points on $y = x^2$ for $x = 1$ to $x = 1.1$, $x = 2$ to $x = 2.5$, $x = 3$ to $x = 3.2$.
4. Draw the graph $y = x^3$ from $x = 0$ to $x = 2$. On it draw the chord joining $(1, 1)$ to $(1.2, 1.73)$ and calculate its gradient. Draw the tangent at $(1, 1)$ and from the graph find the gradient of the curve when $x = 1$.
5. Using the formula for dy/dx, find algebraically the derivative of y with respect to x when $y = x^3$. By substituting calculate the derivatives of $y = x^3$ when $x = -3$, $x = 0$ and $x = 1$.
6. Starch can be hydrolysed (digested) by an enzyme in human saliva; if a starch suspension is mixed with saliva, samples can be removed at intervals and the amount of remaining starch measured, and hence the percentage of starch already hydrolysed determined. The table gives the results for 4 different temperatures (notice that the first sample is taken after 1 minute; earlier samples would be valuable but are technically difficult).

Time/min		1	2	3	4	5	6	7	8	9	10	11	12
% starch hydrolysed	at 15°C	16	36	49	61	70	77	82	87	93	95	98	100
	at 25°C	23	47	64	77	83	88	91	95	98	99	100	100
	at 35°C	42	79	91	94	99	100	100	100	100	100	100	100
	at 45°C	46	89	96	100	100	100	100	100	100	100	100	100

(a) For each temperature plot a graph of % starch hydrolysed against t, showing all 4 curves on one set of axes.
(b) Determine the gradient of the steepest part of the curve for each temperature, stating it in the appropriate units.
(c) The temperature coefficient

$$Q_{10} = \frac{\text{rate at } (T_0 + 10)^\circ\text{C}}{\text{rate at } T_0{}^\circ\text{C}}.$$

Determine Q_{10} for the temperature ranges 15°C to 25°C, 25°C to 35°C, 35°C to 45°C, using the results of (b).

(d) Which 10°C rise in temperature causes the greatest rise in the reaction rate? (Q_{10}, though constant over wide temperature ranges for many chemical reactions, often varies with temperature in biological systems involving enzymes.)

DIFFERENTIATION: SOME RESULTS

	function	*derivative*
(a)	$y = x^n$	$\frac{dy}{dx} = nx^{n-1}$
	$y = ax^n$	$\frac{dy}{dx} = anx^{n-1}$
	$y = a$	$\frac{dy}{dx} = 0$
(b)	$y = ax^p + bx^q$	$\frac{dy}{dx} = apx^{p-1} + bqx^{q-1}$
	$y = af(x) + bg(x)$	$\frac{dy}{dx} = af'(x) + bg'(x)$
(c)	$y = \sin x$ (x in radians)	$\frac{dy}{dx} = \cos x$
	$y = \cos x$	$\frac{dy}{dx} = -\sin x$
(d)	$y = \sin(kx)$ (x in radians)	$\frac{dy}{dx} = k\cos(kx)$
	$y = \cos(kx)$	$\frac{dy}{dx} = -k\sin(kx)$
(e)	$y = e^x$	$\frac{dy}{dx} = e^x$
	$y = e^{kx}$	$\frac{dy}{dx} = ke^{kx}$
(f)	$y = \ln x$ (or $y = \log_e x$)	$\frac{dy}{dx} = \frac{1}{x}$.

We have derived the result for $y = x^n$ in the case where $n = 2$. The result holds for any value of n. Multiplying either x^2 or x^3 by a constant will easily be seen to multiply the derivative by the same constant, and this is true for any differentiable function. The extension to $x^p + x^q$ can also be readily verified.

The other functions differentiated need more skill and are included here because the results are needed; however $y = e^x$ is studied in detail again in chapter 10. A careful look at the graphs of $y = \sin x$ and $y = \cos x$ (figure

7.7) will confirm the results in (c); when differentiating, angles are *always* expressed in radians.

SIMPLE HARMONIC MOTION, S.H.M.

In simple harmonic motion (S.H.M.) the displacement, x of a particle, is given by $x = a \sin \omega t$. Find expressions for the velocity and acceleration of the particle in terms of t, and derive an equation for

$$\frac{d^2x}{dt^2} \text{ in terms of } x\text{, and for } v \text{ in terms of } x.$$

$$x = a \sin \omega t. \tag{1}$$

Differentiating, $\quad$ velocity $v = \dfrac{dx}{dt} = a\omega \cos \omega t. \qquad (2)$

Differentiating again gives

$$\text{Acceleration } f = \frac{dv}{dt} = -a\omega^2 \sin \omega t. \tag{3}$$

Now $\quad v = \dfrac{dx}{dt};$

differentiating v with respect to t,

$$\frac{dy}{dt} = \frac{d}{dt}\left(\frac{dx}{dt}\right) = \frac{d^2x}{dt^2}$$

$$\Rightarrow \frac{d^2x}{dt^2} = -a\omega^2 \sin \omega t. \tag{4}$$

But $a \sin \omega t = x$; substituting,

$$\frac{d^2x}{dt^2} = -\omega^2 x; \tag{5}$$

$\left(\dfrac{d^2x}{dt^2}\right.$ is read as dee two x by dee tee squared.$\left.\right)$

Equation (5) is the *general* equation for S.H.M., equation (1) is the expression for x in the particular case when, at time $t = 0$, $x = 0$. Returning to equation (2), namely $v = a\omega \cos \omega t$, we wish to express v in terms of x. We have

$$x = a \sin \omega t \quad \text{and} \quad v = a\omega \cos \omega t.$$

Now
$$\cos^2 \omega t + \sin^2 \omega t = 1$$
$$\Rightarrow \cos^2 \omega t = 1 - \sin^2 \omega t$$
$$\Rightarrow \cos \omega t = \pm\sqrt{1 - \sin^2 \omega t}$$

and
$$a \cos \omega t = \pm\sqrt{a^2 - a^2 \sin^2 \omega t};$$

but substituting $x = a \sin \omega t$ we have

$$v = a\omega \cos \omega t = \pm\, \omega\sqrt{a^2 - x^2}.$$

MAXIMA AND MINIMA

The graph of $y = a$ where a is constant is a straight line parallel to the x axis, with gradient 0. This is consistent with the result, of course, that when $y = a$, $dy/dx = 0$. Now when a curve reaches a maximum or a minimum value the gradient will at that point be 0 (see figure 8.4) because the tangent there is parallel to the x axis. Notice also that the sign of the gradient changes; for a minimum from negative to positive as x increases, for a maximum from positive to negative.

Fig. 8.4.

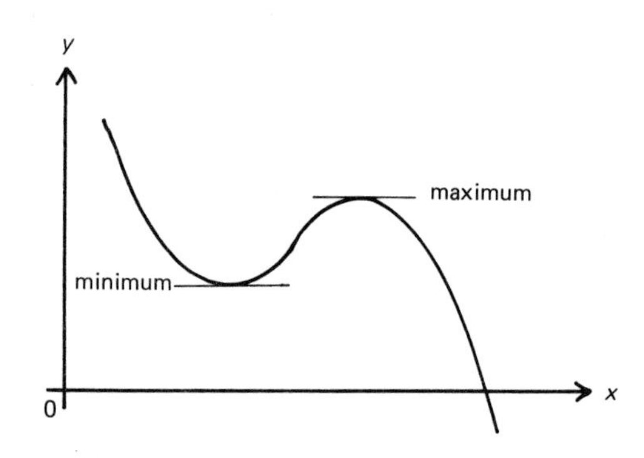

Example 2

Find the value of t for which $s = 90t - 10t^2$ is a minimum or maximum; determine which of the two is found and calculate the corresponding value of s.

$$s = 90t - 10t^2$$

$$\frac{ds}{dt} = 90 - 20t.$$

When $\frac{ds}{dt} = 0$, $\quad 90 - 20t = 0,$

$$t = 4.5.$$

We could draw a graph in the region of $t = 4.5$ to discover the nature of the turning point, but instead find the gradient at $t = 4$ and $t = 5$ (one on each side of $t = 4.5$).

Gradient at $t = 4$ is $90 - (20 \times 4) = +10$.
Gradient at $t = 5$ is $90 - (20 \times 5) = -10$, and so there is a maximum at $t = 4.5$.

$$\begin{aligned} s &= 90t - 10t^2 \\ &= 90 \times 4.5 - 10x(4.5)^2 = 405 - 202.5 = 202.5. \end{aligned}$$

This has been posed as a mathematical problem using t and s instead of x and y, but it could well describe motion, with s in metres, t in seconds and v in ms^{-1}.

In this chapter you have been introduced to the basic ideas of the Calculus, given sufficient technique to attempt some problems, and shown a few applications of the new methods. Exercise B gives you some examples for practice, but for more advanced work you must go to a textbook on the subject.

Exercise B

1. Differentiate $y = x^5$, $y = x^{-3}$, $y = x^{1/2}$.
2. Find the gradient of $y = x^3$ when $x = 2, 0, -2, 5$.
3. Find the gradient of $y = 1/x$ when $x = \frac{1}{2}, 3, -\frac{1}{2}, -3$. Draw a rough sketch of the curve, showing the tangents on it.
4. Differentiate $y = 2x$ and $y = 5$. Explain your results by drawing rough sketch graphs.
5. Differentiate $y = 2\sin x + 3\cos x$.
6. Find the gradient of $y = \cos x$ when $x = 0, \pi/3, \pi/2, \pi/4$.
7. The law of electromagnetic induction can be stated

$$E = \frac{-\mathrm{d}\Phi}{\mathrm{d}t}, \quad \text{where } E \text{ is the induced e.m.f. and } \frac{\mathrm{d}\Phi}{\mathrm{d}t}$$

is the time rate of magnetic flux cutting. When a rectangular coil, area A, rotates with steady angular velocity, ω, about an axis in its own plane, perpendicular to a magnetic flux density B then the flux, Φ, is given by $\Phi = BA\cos\omega t$. Using the law, show that the induced e.m.f. varies sinusoidally with time, that is, $E = E_0 \sin\omega t$. What is the expression for E_0?
8. When $y = 8x - \frac{1}{2}x^2$, find the value of x for which y is a maximum.
9. When $y = a\sin\omega t$, find the first value of t for which y is a minimum (remember that angles are measured in radians in such problems).
10. Find the turning points of $y = 2x^3 - 3x^2 - 36x + 4$; give the values of y at these points and say whether each is a maximum or a minimum.

9

CALCULUS; INTEGRATION

Fig. 9.1.

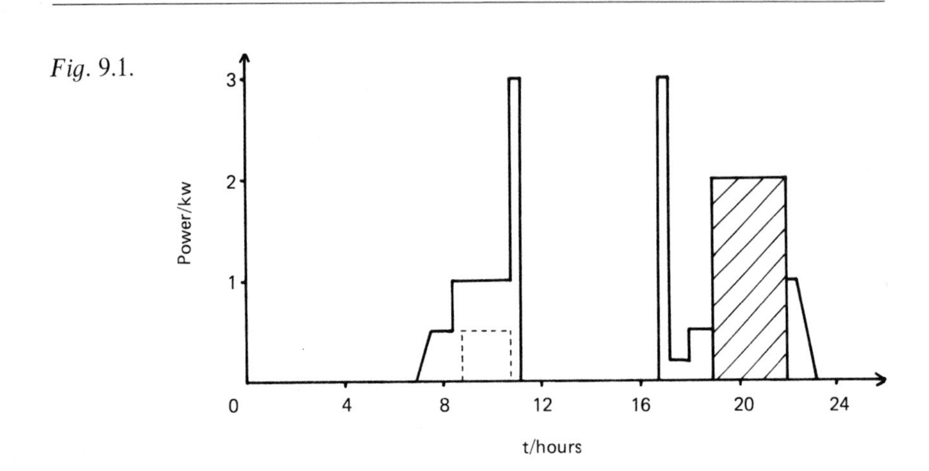

AREA UNDER A GRAPH

How do we measure the consumption of electricity? Drawing a graph showing the consumption in a household over 24 hours should help to answer this question; knowing the wattage of the appliances in use makes this a simple matter, and the result is shown in figure 9.1. Between 19.00 and 22.00 hours the rate of consumption is constant at 2 kW, so that the total energy used in this period is 2×3 kilowatt-hours, 6 kW h in fact. On the graph the shaded area *represents* the energy consumption during this time, showing us that we can calculate this quantity by finding the area between the graph and the time axis over the period of time with which we are concerned. This is a practical proposition, and although no graphs are drawn by the Electricity Boards' meters, the meter effectively performs this summation continuously.

Figure 9.2 illustrates another example of the use of areas on graphs. It shows speed (km h^{-1}) plotted against time (h) for a car journey, and here the area between graph and time-axis for a given period *represents* the distance travelled in that period. In both examples the graph plotted was a rate against time, and the area below the graph represented a physical quantity which itself bears no resemblance to an area.

Now to consider methods of finding such areas; the most straightforward is to count squares. Before you start counting decide on a unit square and calculate its value.

In figure 9.1, an outlined square represents $2 \times 0.5 = 1$ kW h, in figure 9.2 the square represents $0.5 \times 25 = 12.5$ km. Notice, too, that any part of the area below the x axis is taken to be negative. Counting squares soon palls as an entertainment, however, and you get good results by calculation using the trapezium rule.

Fig. 9.2.

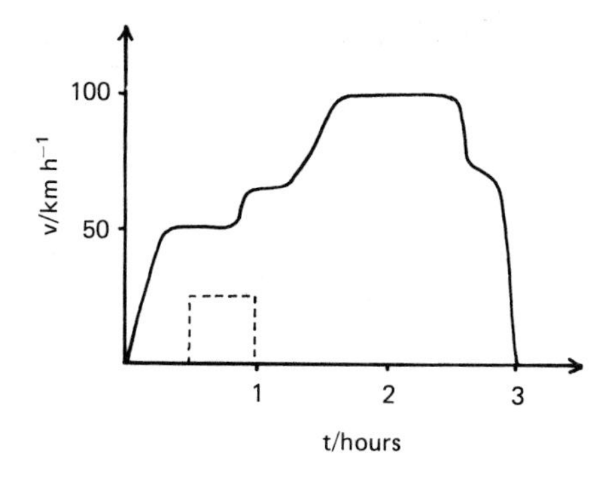

Fig. 9.3.

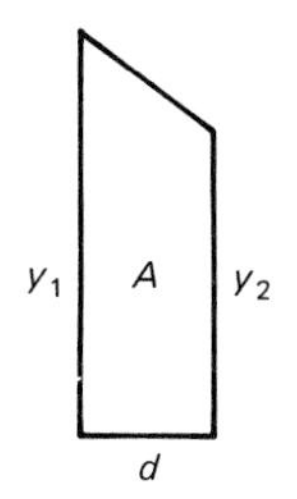

TRAPEZIUM RULE

A simple trapezium, area A, is shown in figure 9.3 with parallel sides y_1 and y_2 and one side at right angles to these. Since

$$\text{area of trapezium} = \begin{pmatrix}\text{distance between}\\ \text{parallel sides}\end{pmatrix} \times \begin{pmatrix}\text{average of}\\ \text{parallel sides}\end{pmatrix}$$

$$A = d \times \left(\frac{y_1 + y_2}{2}\right). \qquad (1)$$

Fig. 9.4.

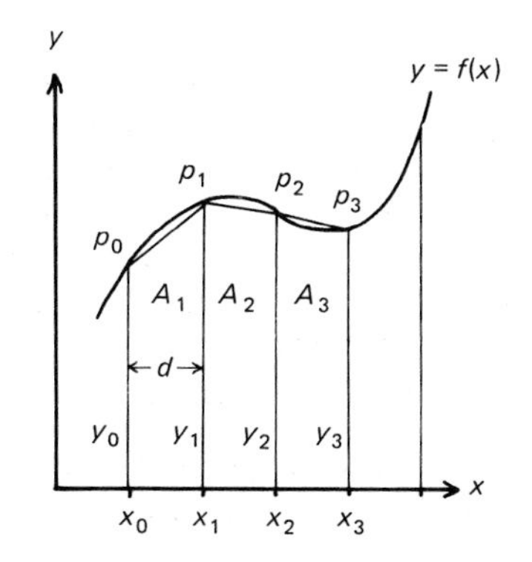

Therefore if we have a table of values of a function $y = f(x)$ for $x_0, x_1 \dots x_n$ we can divide the area beneath the graph of $y = f(x)$ into a series of trapezia, $A_1, A_2, \dots A_n$.

The table that follows gives corresponding values of x and y, and figure 9.4 shows this graphically; all the coordinates are equally spaced (this is essential if the Trapezium Rule is to be used) so that for instance $x_1 - x_0 = d$.

x	x_0	x_1	x_2	$\dots$	x_n
y	y_0	y_1	y_2	$\dots$	y_n

Join P_0P_1, P_1P_2, $\dots$ $P_{n-1}P_n$, and you will see that the whole area has been divided into a series of trapezia $A_1, A_2, \dots A_n$ where, from equation (1),

$$A_1 = \frac{d}{2}(y_0 + y_1),\ A_2 = \frac{d}{2}(y_1 + y_2),\ \dots\ A_n = \frac{d}{2}(y_{n-1} + y_n).$$

$$\begin{aligned}\text{Total area required} &= A_1 + A_2 + \dots + A_n \\ &= \frac{d}{2}\{(y_0 + y_1) + (y_1 + y_2) + \dots + (y_{n-1} + y_n)\} \\ &= d\left\{\frac{(y_0 + y_n)}{2} + (y_1 + y_2 + \dots + y_{n-1})\right\}.\end{aligned}$$

This is often remembered as

$$\text{Area} = d\{(\text{half the ends}) + (\text{the middles})\}.$$

A worked example follows.

Example 1

The table below gives corresponding values of θ and $\sin^2\theta$ $(0 \leqslant \theta \leqslant \pi)$ at intervals of $\pi/12$ radians. Using the trapezium rule calculate the area, A, below the curve.

θ	0	$\frac{\pi}{12}$	$\frac{\pi}{6}$	$\frac{\pi}{4}$	$\frac{\pi}{3}$	$\frac{5\pi}{12}$	$\frac{\pi}{2}$	$\frac{7\pi}{12}$	$\frac{2\pi}{3}$	$\frac{3\pi}{4}$	$\frac{5\pi}{6}$	$\frac{11\pi}{12}$	π
$\sin^2\theta$	0	0.07	0.25	0.50	0.75	0.93	1.00	0.93	0.75	0.50	0.25	0.07	0

Notice first that the values of $\sin^2\theta$ (as you should expect) are symmetrical about $\theta = \pi/2$, so you need only sum from 0 to $\pi/2$. $d = \pi/12$. Keep the "ends" in one column, "middles" in another as we have done in the following calculation to prevent mistakes.

θ	*Ends*	*Middles*	*Calculation*
0	0		
$\frac{\pi}{12}$		0.07	$d = \frac{\pi}{12}$
$\frac{\pi}{6}$		0.25	$\frac{1}{2}A = d\{\frac{1}{2}(\text{ends}) + (\text{middles})\}$
$\frac{\pi}{4}$		0.50	$= \frac{\pi}{12}(0.5 + 2.50)$
$\frac{\pi}{3}$		0.75	$= \frac{\pi}{12}(3.00)$
$\frac{5\pi}{12}$		0.93	$= \frac{\pi}{4}$
$\frac{\pi}{2}$	1.06		$A = \frac{\pi}{2}$

Note that there is no need for a graph although one is given here (figure 9.5) to illustrate the process.

Sometimes, you have the information in the form of a graph, when you read off the y values at equal intervals; sometimes you know the relation (in example 1, $y = \sin^2 \theta$) and can construct your own table. If you have to choose the interval for yourself you have to balance the increased work involved in making d small against the greater accuracy thereby achieved.

There is a more accurate numerical method for obtaining areas known as Simpson's Rule, but the Trapezium Rule should suffice for any application you are likely to encounter.

Fig. 9.5.

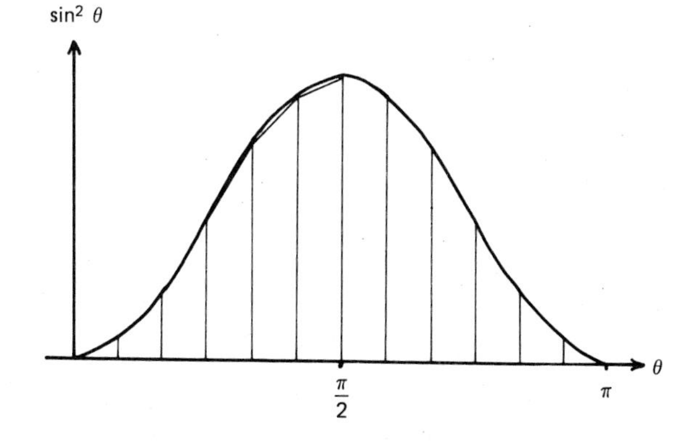

If there is a mathematical model for the phenomenon being studied (that is to say there is a mathematical expression relating two variables such as $pv = \text{constant}$, $s = u + at$, $d\theta/dt = \theta_0 - \theta$) we may expect to find an analytical method for finding the area between curve and axis; from now on we refer to such an area as the area under the curve. The process of calculating the area is called **integration**; what we have done so far is **numerical integration.**

Exercise A

1. For each of the following find the area bounded by the line, the x axis and the ordinates at the given values of x. Illustrate with sketches.
   (i) $y = 11$, from $x = 3$ to $x = 11$,
   (ii) $y = 2x$, from $x = 2$ to $x = 4$,
   (iii) $y = \frac{1}{3}x - 5$ from $x = 7$ to $x = 12$.
2. The speed of a motor-cycle at different times is given in the table below. Use the Trapezium Rule to find the distance travelled by the motor cycle in the first 8 seconds of its motion.

t/s	0	1	2	3	4	5	6	7	8
v/ms^{-1}	0	8.5	16.0	22.5	28.0	32.5	36.0	38.5	40.0

3. Using an interval of $\pi/12$ find the area, between 0 and π, under the curve (a) $y = \sin\theta$, (b) $y = \cos\theta$. Comment on your results, illustrating by sketches.
   Indicate the most economical method of doing the calculation and for (b) give the numerical value of the area required.
4. By finding the values of y at intervals of 0.5 and using the Trapezium Rule find the area under the curve $y = e^x$ between $x = -2.0$ and $x = 2.0$.
5. When a gas at pressure P expands by a small volume ΔV then the work done W is given by the expression

$$W = P \times \Delta V$$

   assuming the pressure remains constant during the small change in volume. Hence by calculating the area between the P against V curve and the V axis, it is possible to find the work done on or by a gas.
   (a) For a particular gas undergoing an isothermal (i.e. constant temperature) expansion $PV = 10^3$ energy units. Use the Trapezium Rule to find the work done by the gas in expanding from an original volume $V_0 = 10\,\mathrm{cm}^3$ to a final volume $V_f = 20\,\mathrm{cm}^3$. Draw a graph of P against V from $V_0 = 10\,\mathrm{cm}^3$ to $V_f = 20\,\mathrm{cm}^3$ and use it to help you decide on a suitable width for the trapezia.
   (b) This gas is now compressed adiabatically from $V_f = 20\,\mathrm{cm}^3$ to the original volume $V_0 = 10\,\mathrm{cm}^3$. For this adiabatic change, $PV^{1.4} = 3314$ energy units. Add corresponding values of P and V to the graph in (a). Use the Trapezium Rule to calculate the work done on the gas during this compression.
   Hence find the difference between the work done by the gas in (a) and the work done on the gas in (b). Use your graph to make a quick approximate check on this result, explaining your reasoning.

Fig. 9.6.

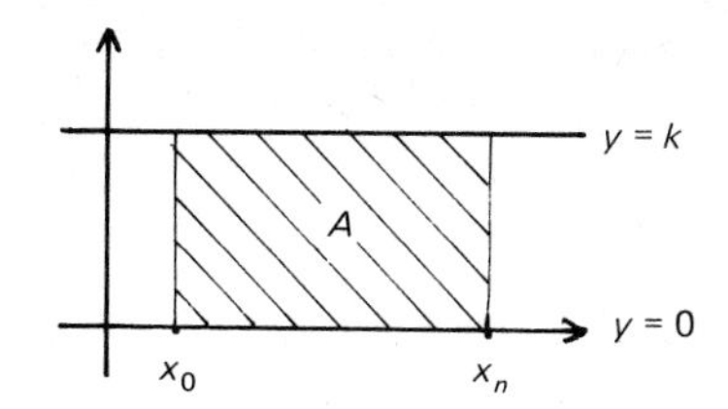

INTEGRATION; $y = c, y = mx + c$.

We begin by calculating areas of regions bounded by straight lines. A pattern should emerge, and we can then extend our investigation to regions of greater complexity. The simplest case of all is illustrated in figure 9.6. A is the area bounded by $y = k$, $y = 0$, $x = x_n$ and $x = x_0$, and since A is a rectangle

$$A = k(x_n - x_0).$$

Next suppose the boundaries of A are $y = x$, $y = 0$ and $x = x_n$ (figure 9.7). A is a triangle, base x_n, height x_n (since $y = x$ for all x) so that

$$\begin{aligned} A &= \tfrac{1}{2}(x_n \times x_n) \\ &= \tfrac{1}{2}x_n^2. \end{aligned}$$

We might require the area bounded by $y = x$, $y = 0$, $x = x_n$ and $x = x_0$, that is to say the area under $y = x$ between $x = x_0$ and $x = x_n$. From figure 9.8 this is a trapezium, but it will be better to think of it as the difference in

Fig. 9.7.

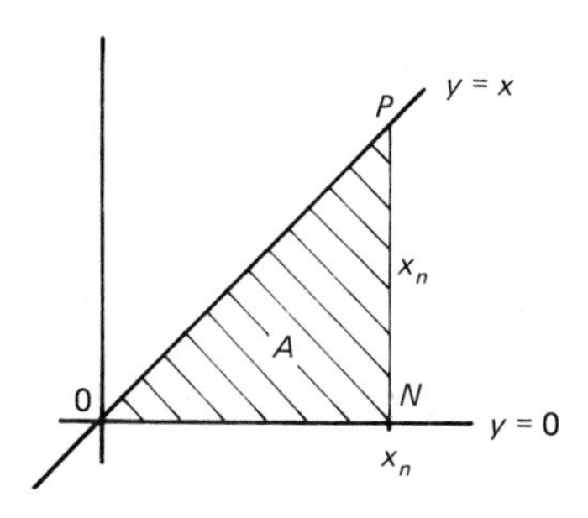

Fig. 9.8.

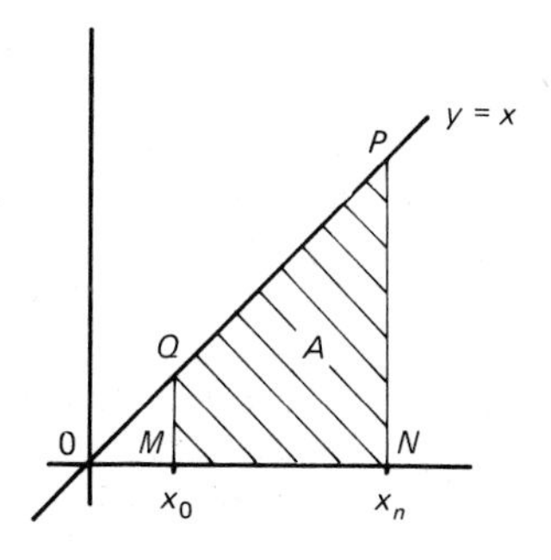

area of two triangles so that

$$A = \text{area } OPN - \text{area } OQM$$

$$A = \tfrac{1}{2}x_n^2 - \tfrac{1}{2}x_0^2 = \tfrac{1}{2}(x_n^2 - x_0^2).$$

If, instead of $y = x$ the boundary were $y = mx$, when $x = x_0$, $y = mx_0$, and

$$A = \tfrac{1}{2}x_n \times mx_n = \tfrac{1}{2}mx_n^2 \text{ (figure 9.9)},$$

and if the area required lies between $y = mx$, the x axis and the ordinates x_0 and x_n,

$$\begin{aligned} A &= \tfrac{1}{2}(mx_n^2 - mx_0^2) \\ &= \tfrac{1}{2}m(x_n^2 - x_0^2). \end{aligned}$$

One more example, and the case for straight line boundaries is complete; let $y = mx + c$ be the boundary (figure 9.10). The line $y = c$ is shown, and the area A is seen to consist of a triangle, A_1 and a rectangle A_2.

$$\begin{aligned} A &= A_1 + A_2 \\ &= \tfrac{1}{2}m\,x_n^2 + cx_n. \end{aligned}$$

Fig. 9.9.

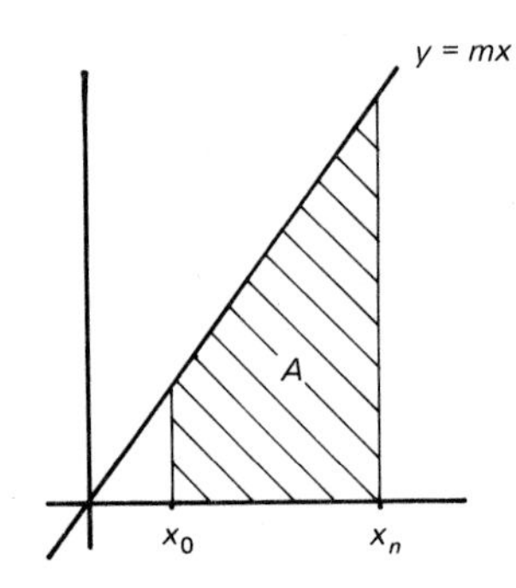

Fig. 9.10.

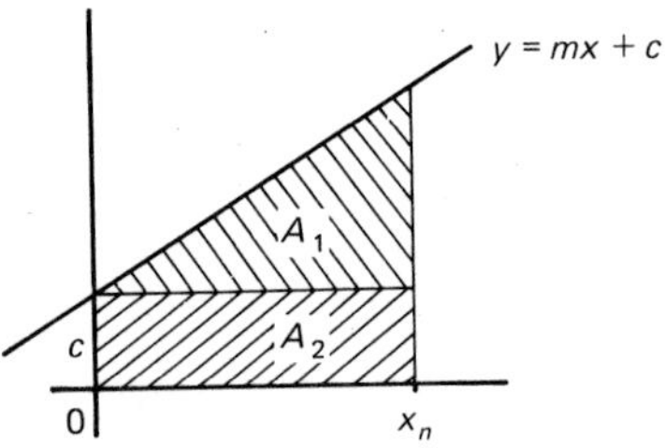

If the area lies between x_0 and x_n

$$\begin{aligned} A &= \tfrac{1}{2}m(x_n^2 - x_0^2) + c(x_n - x_0) \\ &= \tfrac{1}{2}mx_n^2 + cx_n - (\tfrac{1}{2}mx_0^2 + cx_0) \\ &= [\tfrac{1}{2}mx^2 + cx]_{x_0}^{x_n}. \end{aligned}$$

This is the standard notation for this area; square brackets are always used and x_0 and x_n are called the **lower and upper limits of integration** respectively. It is clear, concise and universal, and to help you to become accustomed to it here are the results we have obtained restated in this form.

Boundary curve $y = f(x)$	A, Area under the curve between x_0 and x_n
$y = c$, c constant	$[cx]_{x_0}^{x_n}$
$y = mx$, m constant	$\frac{1}{2}[mx^2]_{x_0}^{x_n}$
$y = mx + c$	$[\frac{1}{2}mx^2 + cx]_{x_0}^{x_n}$

The section which follows illustrates the use of these techniques.

Fig. 9.11.

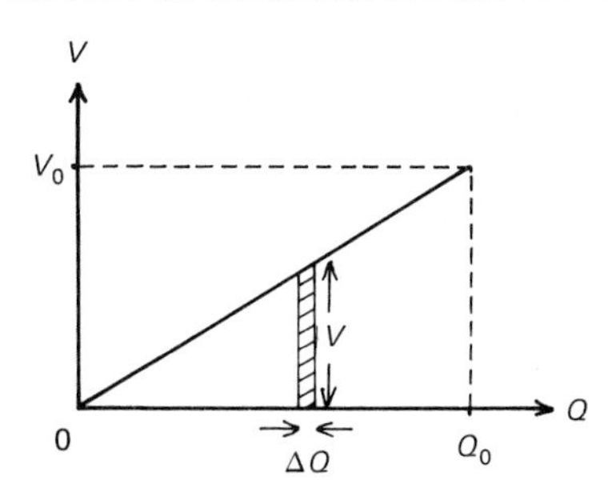

DISCHARGE OF A CAPACITOR

For a capacitor $V \propto Q$ where V is the p.d. and Q the charge. This is shown graphically by the straight line in figure 9.11, V_0 being the p.d. initially, with corresponding charge Q_0.

Suppose the p.d. is V when the capacitor starts to discharge, a small quantity ΔQ of charge passing between the plates. Then the work done (or energy lost) W is represented by the area at the shaded strip in figure 9.11. If ΔQ is small we can assume that the p.d. remains constant, and

$$W = V\Delta Q.$$

If the capacitor discharges completely from $V = V_0$ to $V = 0$ the loss of energy is represented by the sum of all such strips between $Q = 0$ and $Q = Q_0$. From figure 9.11 we see that this sum is a triangle with base Q_0, height V_0,

$$W = \tfrac{1}{2} Q_0 V_0. \qquad (1)$$

The actual relation between V and Q is

$V = Q/C$ where C (a constant) is the capacitance.

Thus we can think of the area representing the work as the area under the line $V = Q/C$ between $Q = 0$ and $Q = Q_0$, written

$$W = \left[\frac{1}{2}\frac{Q^2}{C}\right]_0^{Q_0} = \frac{1}{2}\frac{Q_0^2}{C}. \qquad (2)$$

Referring once more to figure 9.11 you will see that the triangle bounded by the line $Q = CV$, the V axis and the line $V = V_0$ is equal in area to the triangle whose area we have already calculated. Hence also

$$W = \left[\frac{1}{2} C V^2\right]_0^{V_0} = \frac{1}{2} C V_0^2. \tag{3}$$

In this example the calculation of area is simple and (2) and (3) can be derived directly by substituting in equation (1).

Fig. 9.12.

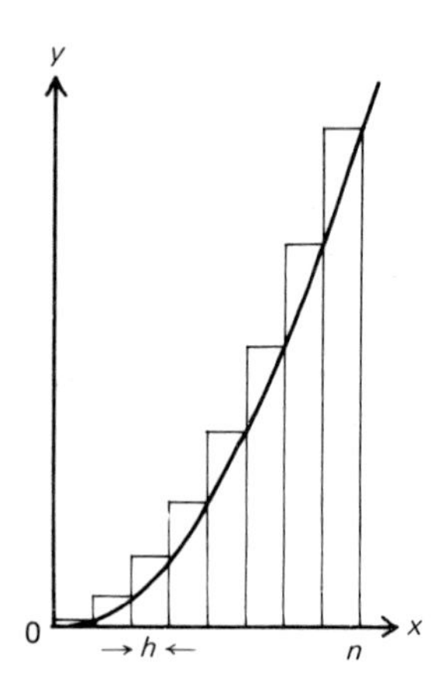

INTEGRATION: $y = x^2$

The smooth curve in figure 9.12 is the graph of $y = x^2$; finding the area below the curve between $x = 0$ and $x = n$ requires a new technique. The rectangles in figure 9.12 all have the same width h, and the sum of their areas provides an approximation to the area we want. As h gets smaller the sum of the rectangles more nearly approaches the area under the curve until, in the limit, when $h \to 0$, we have an expression for the required area.

In figure 9.12 there are 8 rectangles $\Rightarrow n = 8h$, or $h = n/8$; the table gives corresponding values of x and y. We need also the result that

$$\sum_{i=1}^{k} i^2 = (1/6)\,k\,(k+1)\,(2k+1).$$

x	0	$1h$	$2h$	$3h$	$4h$	$5h$	$6h$	$7h$	$8h$
y	0	h^2	$(2h)^2$	$(3h)^2$	$(4h)^2$	$(5h)^2$	$(6h)^2$	$(7h)^2$	$(8h)^2$

From the table we see that

$$\text{sum of areas of rectangles} = A = h \times h^2 + h \times (2h)^2 + \ldots + h \times (8h)^2$$

$$= h^3(1^2 + 2^2 + \ldots + 8^2) = h^3 \sum_{i=1}^{k} i^2.$$

$$\text{But } h = n/8 \Rightarrow A = (n/8)^3\,(1/6)8(8+1)(16+1)$$

$$= (n^3/6)(8/8)\{(8+1)/8\}\{(18+1)/8\},$$

$$\Rightarrow A = (n^3/6)\,(1 + 1/8)\,(2 + 1/8).$$

As $h \to 0$ the number of rectangles, 8 in the figure, increases, and in the brackets the fraction, 1/8 in our calculation, becomes smaller and smaller, until in the limit we have the result

$$\text{(Area under the curve between } x = 0 \text{ and } x = n) = \lim_{h \to 0} A = 2n^3/6$$
$$= n^3/3.$$

The time has come to introduce the **integral sign** $\int$ (it is a modification of the letter S) and to write the result

$$\int_0^n x^2 \, dx = \left[\frac{x^3}{3} \right]_0^n = \frac{n^3}{3}, \text{ or in words,}$$

the integral of x^2 *with respect to* x, between the limits 0 and n, is $n^3/3$. Figure 9.13 illustrates the integration of a function $f(x)$. The area under the graph is the limit of the sum of rectangles, length $f(x)$, width δx, when δx tends to 0, written

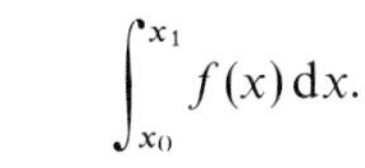

$$\int_{x_0}^{x_1} f(x) \, dx.$$

Fig. 9.13.

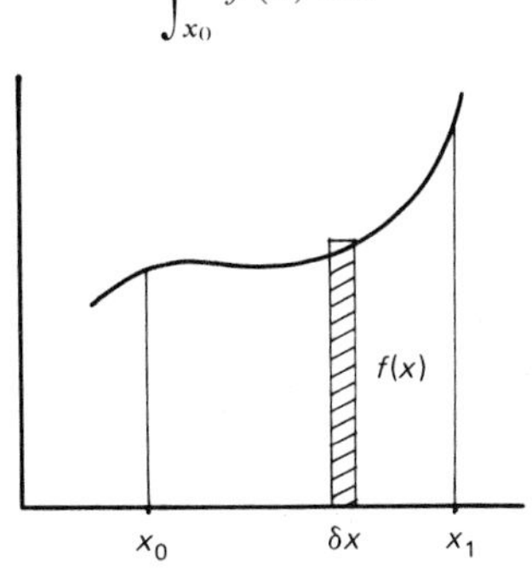

FURTHER INTEGRATION

It is possible to treat $y = x^3$, $y = x^4$ as before, but the algebra becomes increasingly difficult. Besides, there are many other functions which we wish to integrate, and a much more fruitful approach is suggested by the following comparisons:

Integrating	Differentiating
$\int_0^n k dx = kn$	$\frac{d}{dx}(kx) = k$
$\int_0^n x dx = \frac{n^2}{2}$	$\frac{d}{dx}\left(\frac{x^2}{2}\right) = x$
$\int_0^n x^2 dx = \frac{n^3}{3}$	$\frac{d}{dx}\left(\frac{x^3}{3}\right) = x^2$

The above results suggest that **integration is the inverse operation to differentiation**, and in general this is true. Henceforth you can use this knowledge when integrating, and as a check, differentiate the result of your integration. It follows at once that

$$\int_{x_1}^{x_2} x^n \, dx = \left[\frac{x^{n+1}}{n+1} \right]_{x_1}^{x_2},$$

for all values of n except $n = -1$.

When $n = -1$ i.e. $x^{-1} = 1/x$, we have the special case

$$\int_{x_1}^{x_2} (1/x) \, dx = [\log_e x]_{x_1}^{x_2}$$

$$= \log_e x_2 - \log_e x_1$$
$$= \log_e (x_2/x_1).$$

We give the result here for reference, that

$$\int_{x_1}^{x_2} e^{kx} dx = [(1/k) \, e^{kx}]_{x_1}^{x_2},$$

although the exponential function is considered in more detail in chapter 10.

$$\int_{\theta_1}^{\theta_2} \sin k\theta \, d\theta = [-(1/k) \cos k\theta]_{\theta_1}^{\theta_2}$$

and

$$\int_{\theta_1}^{\theta_2} \cos k\theta \, d\theta = [(1/k) \sin k\theta]_{\theta_1}^{\theta_2};$$

check these for yourself by differentiating the results. Integration of $\tan \theta$ requires a little more technique than you yet have, but is given here for reference:

$$\int_{\theta_1}^{\theta_2} \tan k\theta \, d\theta = [(1/k) \ln |\sec k\theta|]_{\theta_1}^{\theta_2}.$$

Another trigonometrical integral often needed is

$$\int_{\theta_1}^{\theta_2} \sin^2 \theta d\theta = \left[\frac{1}{2} \left(\theta - \frac{\sin 2\theta}{2} \right) \right]_{\theta_1}^{\theta_2}.$$

There is a great deal more to integration than this, many more standard integrals, ingenious methods enabling you to integrate very complicated functions. There are lists in reference books (a comprehensive one in S.M.P. Advanced Tables, Cambridge University Press, Second Edition 1971).

We show in example 2 how integration is used to solve a problem in electricity, and give some simple problems in exercise B.

Example 2

Find the root-mean-square value of an alternating current I whose variation with time t can be represented by the expression $I = I_0 \sin \omega t$ where I_0 is the maximum current and $\omega = 2\pi f$, f being the number of cycles of alternation per second.

The root-mean-square (r.m.s.) value of an alternating current is defined as the square root of the mean (or average) value of the squares of the current during one cycle. Thus

$$I_{\text{r.m.s.}} = \sqrt{\text{mean value of } I_0^2 \sin^2 \omega t}$$
$$= I_0\sqrt{\text{mean value of } \sin^2 \omega t.} \qquad (1)$$

Figure 9.14 shows the variation of $\sin \omega t$ and $\sin^2 \omega t$ with ωt. The area of the shaded strip is given by $\sin^2 \omega t(\omega\delta t)$, where δt is very small. Put $\omega t = \theta$; then differentiating, $\omega(\mathrm{d}t/\mathrm{d}\theta) = 1 \Rightarrow \omega\delta t \quad \delta\theta$, and we have $\sin^2 \omega t(\omega\delta t) \approx \sin^2 \theta\delta\theta$ and we can make a direct comparison with example 1, where the total area A between the curve $\sin^2 \theta$ and the θ axis (see figure 9.5) was calculated using the trapezium rule and found to be $\pi/2$.

Then the total area for one cycle is $A' = \pi$.

Thus the mean value of $\sin^2 \omega t$ over a complete cycle (i.e. from $\omega t = 0$ to $\omega t = 2\pi$) is given by

$$A/(\text{total value of } \omega t \text{ over one cycle}) = \pi/(2\pi) = 1/2.$$

Substituting in (1) we obtain

$$I_{\text{r.m.s.}} = I_0\sqrt{(1/2)} = I_0/\sqrt{2} = 0.707\, I_0.$$

Written in integral form the mean value of $\sin^2 \theta$ over a complete cycle is given by

$$\frac{\int_0^{2\pi} \sin^2 \theta \mathrm{d}\theta}{\int_0^{2\pi} \mathrm{d}\theta} = \frac{\frac{1}{2}\left[\theta - \frac{\sin 2\theta}{2}\right]_0^{2\pi}}{[\theta]_0^{2\pi}} = \frac{\pi}{2\pi} = \frac{1}{2}.$$

Fig. 9.14.

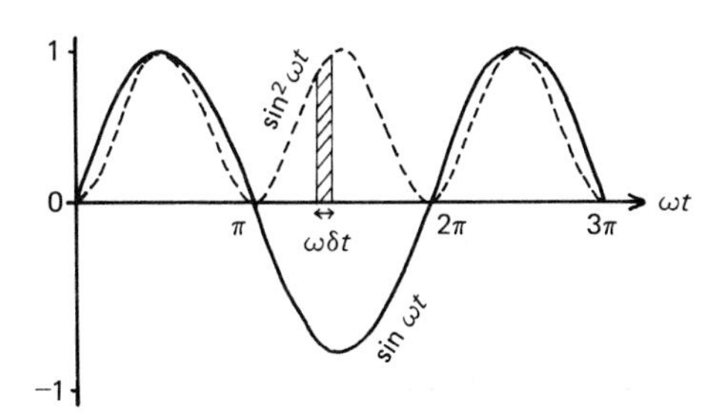

DEFINITE AND INDEFINITE INTEGRALS

Whenever the limits of integration are known the value of the integral can be found exactly, hence the name **definite integral** for every integral given so far in this chapter. When the limits are given in general terms (x_0 and x_1, or θ_1 and θ_2) there is really no less information conveyed if we write, for instance,

$$\int x^3\, \mathrm{d}x = x^4/4 + \text{constant},$$

instead of,

$$\int_{x_1}^{x_2} x^3\, dx = \left[\frac{x^4}{4}\right]_{x_1}^{x_2}.$$

The form of the integral, without limits given, is called an **indefinite integral** and the result must always include a constant (written k or c or a, or any other suitable letter). Then if you get more information you can find the constant; if not, you know the general form of that particular integral.

Exercise B

1. Evaluate

(i) $$\left[\frac{3x^2}{2} - 6x + 5\right]_1^5,$$

(ii) $$\left[\frac{t^3}{3}\right]_{-3}^4,$$

(iii) $$\left[\frac{1}{x^3}\right]_2^3,$$

(iv) $$[2\theta]_0^{\pi/2},$$

(v) $$[\sin\theta]_{\pi/4}^{\pi/2},$$

(vi) $$[\log_e y]_3^5.$$

2. For each of the following show in a rough sketch the area to be determined and work out the integral

(i) $$\int_{-2}^{4} (3 + 2x)\, dx,$$

(ii) $$\int_0^5 \frac{1}{2}x^2\, dx,$$

(iii) $$\int_0^{\pi/4} \sin\theta\, d\theta,$$

(iv) $$\int_0^{\pi} \cos\theta\, d\theta,$$

(v) $$\int_{1/2}^{2} \frac{dx}{x^2},$$

(vi) $$\int_{\pi/6}^{\pi/3} \tan\theta\, d\theta,$$

(vii) $$\int_0^2 e^x\, dx,$$

(viii) $$\int_1^3 e^{-2t}\, dt,$$

(ix) $$\int_1^4 \frac{dx}{x}.$$

3. Perform the following integrations:

(i) $$\int_{-1}^{2} (2x + 3)\,dx,$$

(ii) $$\int_0^5 (x^2 - 2x + 1)\,dx,$$

(iii) $$\int_1^3 \frac{1}{x^3}\,dx,$$

(iv) $$\int_{\pi/6}^{\pi/4} (\sin\theta + \cos\theta)\,d\theta,$$

(v) $$\int_0^{\pi/4} \sin^2\theta\,d\theta$$

(use the result given at the end of example 2).

4. For a stretched spring $F = ke$ where F is the extending force which produces an extension e (k constant).

Show that the area between the line $F = ke$ and the e axis is given by the expression $(1/2)\,Fe$. Hence, using the fact that

work done = (force) × (distance moved in direction of force)

show that the expression $(1/2)\,Fe$ is the work done on the spring in stretching, i.e. the energy stored in the stretched spring.

5. Evaluate the following indefinite integrals:

(i) $$\int (x^3 - 3x)\,dx,$$

(ii) $$\int (3\cos x + 1)\,dx,$$

(iii) $$\int \cos\omega t\,dt.$$

6. Distance travelled, s (metres) in time t (seconds) is given by

$$\int_0^t v\,dt$$

where the speed is v ms^{-1}.

If $v = 9t - \frac{1}{2}t^2$ find the distance travelled in the first 8 seconds of the motion.

10

EXPONENTIAL VARIATION

RADIOACTIVE DECAY

Some mathematical treatment of exponential growth and decay can be found in chapter 3. Here we study a physical phenomenon, radioactive decay, and explore this type of variation from an experimental standpoint, using graphical methods.

Radioactive decay is a completely random process in which nuclei disintegrate quite independently. In a radioactive material where there is a large number of unstable nuclei, it is found that a particular fraction of them always decays in the same length of time, although any one nucleus may last for any length of time. The fraction usually chosen is 1/2, and the time for half the population to disintegrate is called the **half-life**, $T_{1/2}$. There is nothing paradoxical about finding a numerical constant to characterise each substance; the laws of probability apply *because* the process is random. Growth processes show a similar regular pattern of behaviour and there is a wide range of both growth and decay processes in physics, chemistry, biology and economics. All can be described by the same type of equation and investigated either algebraically or graphically.

Example 1

Figure 10.1 shows the variation of a population of radioactive atoms with time for a particular material. Find the half-life $T_{1/2}$ for this substance and an expression for the number of atoms present at time t.

To find $T_{1/2}$ take any value N of the population and from the graph find the time taken for the population to fall to $N/2$. If you do this for several values of N you will find that $T_{1/2} \approx 7$ s. In the table below you will see that $T_{1/2}$ is used as the unit of time.

k	0	1	2	3	4
Population $N/10^6$ atoms	10	5	2.5	1.25	0.625
Time t/s	0	7	14	21	28
Time $T/T_{1/2}$ units	0	1	2	3	4
N_k/N_{k-1}	—	5/10 = 1/2	2.5/5 = 1/2	1.25/2.5 = 1/2	0.625/1.25 = 1/2

The fraction N_k/N_{k-1} is undefined for $k < 1$ unless data for $t < 0$ is available.

We see that $N_k : N_{k-1} = 1 : 2$, i.e. the ratio of the population at the end of an interval of time $T_{1/2}$ to the population at its beginning is a constant, namely 1:2, and the curve shown in figure 10.1 is therefore called a **Constant Ratio Curve** (you will have recognised it as exponential).

Fig. 10.1.

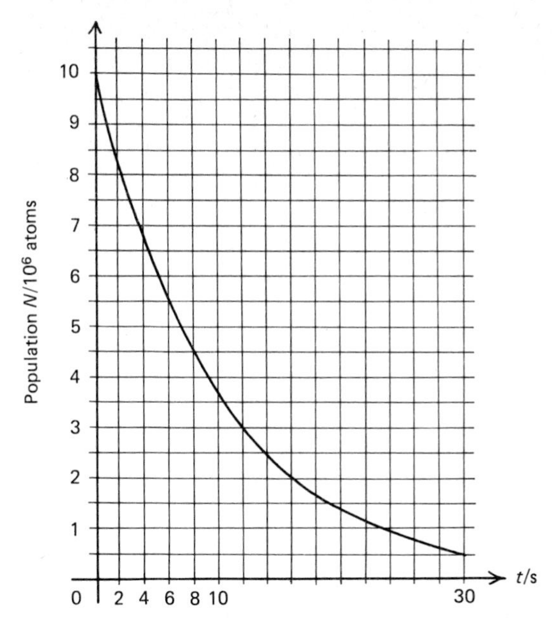

So far we have considered only the ratio 1:2 and its related $T_{1/2}$, but any other ratio could be used. In this example for instance $T_{1/3} = 11$ s for the ratio 1:3.

N_n EXPRESSED IN TERMS OF t AND $T_{1/2}$

Let N_0 be the number of the original population, N_1 the number after 1 half-life, N_2 the number after 2 half-lives and so on. From the table we see that

$$\frac{N_1}{N_0} = \left(\frac{1}{2}\right)^1, \quad \frac{N_2}{N_0} = \left(\frac{1}{2}\right)^2, \quad \frac{N_3}{N_0} = \left(\frac{1}{2}\right)^3,$$

and deduce that

$$\frac{N_n}{N_0} = \left(\frac{1}{2}\right)^n.$$

At time t, $t = nT_{1/2} \Rightarrow n = t/T_{1/2}$, and we have

$$\frac{N_n}{N_0} = \left(\frac{1}{2}\right)^{t/T_{1/2}}$$

or

$$\frac{N_n}{N_0} = 2^{-t/T_{1/2}}$$

$$N_n = N_0 2^{-t/T_{1/2}},$$

which is an exponential variation.

Substituting the values found in example 1 we find that we can express N for that material as

$$N = (10 \times 2^{-t/7}) \times 10^6 \text{ atoms at time } t \text{ seconds.}$$

CONSTANT RATIO CURVES

So far in one example we have studied the ratio 1:2 in detail, and touched on 1:3, and then developed the general expression.

The ratio 1:e has special importance in theoretical work. To remind you once more, e is the irrational number 2.718... ; we have already met e as the base of Napierian logarithms and as these are the logarithms which appear in applications of the Calculus the re-appearance of e should not be too surprising.

The table shows $T_{1/2}$, $T_{1/3}$ and $T_{1/e}$ for the material of example 1.

r	2	3	e
Decay ratio $1/r$	1/2	1/3	$1/e$
$T_{1/r}$/s	7	11	10.2
N/N_0	$2^{-(t/T_{1/2})}$	$3^{-(t/T_{1/3})}$	$e^{-(t/T_{1/e})}$

The expression

$$N = N_0 e^{-t/T_{1/e}}$$

is clumsy but occurs frequently, so it has been simplified to

$$N = N_0 e^{-\lambda t}$$

where $\lambda = 1/T_{1/e}$, called the **decay constant λ** (Greek letter lambda). It should now be clear that the variation of population with time is exponential as this equation is of the same form as $y = a^{-x}$ (see chapter 3).

EXP NOTATION

The use of exp instead of e is widespread and convenient; we have used e so far but instead of, for example, $N = N_0 e^{-\lambda t}$, we can write $N = N_0 \exp(-\lambda t)$. Note that exp is never used for powers of any number other than e.

EXPONENTIAL GROWTH

When a population is growing, not decaying, exponentially the constant ratios to be considered will be n:1, $n > 1$ with corresponding n-life T_n. In exercise A number 1 you are asked to find the double-life T_2, and also T_e.

Exercise A

1. The following table gives the increase in size of a population of aphids.

Time t/day	0	5	10	15
No. of population N	100	305	852	2497

Draw a graph to show these data (very accurate drawing is needed) and from it find the double-life T_2. Using this value, copy and complete the following table from your graph.

$k = t/T_2$	0	1	2	3	4
Time t/day					
Pop. number at time T_k	100				
N_k/N_{k-1}					

From the graph determine T_e and by substituting in $N = N_0 \exp(t/T_e)$ obtain an expression for N for this population.

2. In the germination of a barley seedling, each successive leaf is longer than the previous ones, so that a graph of total height against time has a series of "steps". Plot a graph of the following data:

Day since emerging from soil	0	1	2	3	4	5	6	7	8	9	10	11	12
Height above soil level/cm	0	23	46	68	79	83	83	86	98	108	143	161	170

Day since emerging from soil	13	14	15	16	17	18	19	20	21	22	23	24
Height above soil level/cm	173	176	181	216	267	293	315	320	321	321	322	322

Despite its difference from a typical growth curve, confirm that the increase in height is approximately exponential by calculating the following ratios:

$$\text{(a)} \quad \frac{\text{height at day 14}}{\text{height at day 7}},$$

$$\text{(b)} \quad \frac{\text{height at day 21}}{\text{height at day 14}}.$$

3.

Time t/hour	0	1	2	3	4	5	6	7	8	9	10	11
Population size N	94	125	165	820	287	362	440	520	608	690	760	826
Time t/hour		12	13	14	15	16	17	18	19	20		
Population size N		875	914	941	965	984	995	1005	1010	1011		

(a) The table above shows the growth of a population of bacteria, with a limited food supply, over a period of 20 hours. Plot a graph of absolute growth against time, which will be a typical S-shaped curve. Use the gradient to determine which 3-hour period shows the greatest increase in numbers.

(b) Calculate the hourly increments ΔN, i.e. the increase in numbers between successive readings (see note below). Using your answer to (a), predict the shape of a graph in which these increments are plotted against time. Now plot such a graph, using the same scale as before, on both axes.

(c) Express each hourly increment as a percentage of the population size at the previous count, i.e. express $\Delta N/N$ as a percentage (see note below). Plot these figures against time.

The graph confirms that the small increments at the end do indeed represent a small percentage of the total population size. But notice that, although the actual increments at the beginning are also small, the percentage increase here is greatest. How do you interpret the gradient ≈ 0 over the first few hours?

Note: The following table illustrates a good arrangement for the calculation of ΔN required in (b), and $\Delta N/N$ as a percentage in (c).

t	N	ΔN	$\Delta N/N$	%
0	94			
		31	31/94	33
1	125			
		40	40/125	32
2	165			
		55	55/165	33
3	220			

SUMMARY

Exponential curves are constant ratio curves, that is, in equal time intervals the number of the population, N, changes by a constant ratio. T_n represents the time taken for N to increase itself n times and occurs in a growth curve. $T_{1/n}$ represents the time to reduce to $(1/n)N$ (a decay curve).

Special cases of $T_{1/n}$ are $T_{1/2}$ (half life) and $T_{1/e}$ ($e \approx 2.718$, and $1/e \approx 0.368$).

The decay constant is $\lambda = 1/T_{1/e}$.

The corresponding constant for a growth curve is given by $1/T_e$.

ALGEBRAIC EXPRESSIONS FOR THESE RELATIONS

Decay	*Growth*
$N/N_0 = 2^{-t/T_{1/2}}$	$N/N_0 = 2^{t/T_2}$
$N/N_0 = e^{-\lambda t} = \exp(-\lambda t)$	$N/N_0 = e^{kt} = \exp(kt)$ where $k = 1/T_e$.

RATES OF CHANGE (GRADIENTS)

In a radioactivity experiment it is impossible to measure the number of the population, N, at any instant, but you can find the rate of change of N by recording the number of counts in unit time of a Geiger counter. From this measured rate of change, $\Delta N/\Delta t$, we can derive the mathematical expression for N, and $T_{1/2}$ or λ for the material. We adopt an experimental approach once more, by examining the radioactive material of example 1 again.

Example 2

The radioactive material of example 1 was found to have a half-life $T_{1/2} = 7$ s. From the graph of $N = (10 \times 2^{-t/7}) \times 10^6$ atoms, find $\Delta N/\Delta t$ at $t = 0, 7, 14, 21$ s by drawing tangents. Record these results and for each calculate also $(\Delta N/\Delta t)/N$.

The graph of figure 10.1 is repeated in figure 10.2 which shows the tangents at $t = 0$, 7, 14, 21 s. It is difficult to draw these accurately (see chapter 7), so we call their gradients $\Delta N/\Delta t$ and do not assume that they are the derivatives, dN/dt. We hope however to find clues as to the form of dN/dt from a study of the following table which summarises results derived from figure 10.2.

Fig. 10.2.

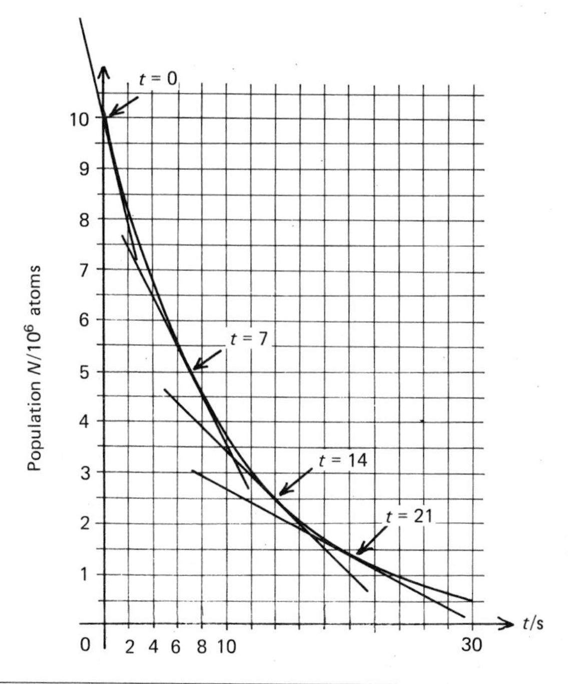

No. of half-lives from $t = 0$	0	1	2	3
time t/s	0	7	14	21
$(\Delta N/\Delta t)/10^6\ s^{-1}$	1	0.50	0.24	0.125
$(\Delta N/\Delta t)_k/(\Delta N/\Delta t)_{k-1}$	—	$\frac{1}{2}$	$\frac{1}{2}$	$\frac{1}{2}$
$(\Delta N/\Delta t)/N/s^{-1}$	0.1	0.1	0.1	0.1

In the 4th row the constant ratio $\frac{1}{2}$ appears again, and the 5th row suggests that we may have $(\Delta N/\Delta t)/N = 0.1\,s^{-1}$, wherever on the curve this fraction is calculated. We must suspect then that the gradient is proportional to N everywhere on the curve, and it is true that the gradient also varies exponentially. All this is experimental, but it can be shown mathematically that differentiating $N = N_0 e^{-\lambda t}$ with respect to time gives

$$\frac{dN}{dt} = -\lambda N, \quad \text{or} \quad \frac{dN}{dt} = -\lambda N_0 e^{-\lambda t}.$$

This enables us to find the derivative given the law for N, and if we know the rate of change, to find T.

RELATION BETWEEN λ AND $T_{1/2}$

We know that $\lambda = 1/T_{1/e}$, but it is the half-life which is more commonly specified, so you need to know the relation between $T_{1/2}$ and λ.

After one half-life $N/N_0 = \frac{1}{2}$, but $N/N_0 = e^{-\lambda t}$ so that for $t = T_{1/2}$ we have

$$\frac{1}{2} = e^{-\lambda T_{1/2}}$$
$$\frac{1}{2} = 1/e^{\lambda T_{1/2}}$$
$$2 = e^{\lambda T_{1/2}}$$

Taking logarithms to the base 10 of both sides of this equation we have

$$\log_{10} 2 = \lambda T_{1/2} \log_{10} e$$
$$\lambda T_{1/2} = (\log_{10} 2)/(\log_{10} e)$$
$$= 0.3010/0.4343$$
$$= 0.693$$
$$\lambda = 0.693/T_{1/2}$$

In example 1 $T_{1/2} = 7\,s$ which gives $\lambda = 0.099$.

For a growth curve, when we have to consider T_2 the **double-life**,

$$N/N_0 = 2 \quad \text{and} \quad N/N_0 = e^{kT_2}$$

so that the $\log_{10} 2 = kT_2 \log_{10} e$, the same relation as before.

TESTS FOR EXPONENTIAL VARIATION

CONSTANT RATIO

Apply the techniques of examples 1 and 2, looking for the constant ratio pattern. This means drawing a graph, tabulating results from it and from this tabulation deducing the relation.

LOG GRAPH

This has the advantage of being a straight line if the relation is exponential (you may remember a similar procedure used to test for $y = kx^n$ which gives a linear log-log graph). The reason for this follows, and then a description of the procedure in case the mathematical argument does not interest you very much.

Suppose that $N = N_0 e^{-\lambda t}$; take logs of both sides (the base does not matter so we use common logs for convenience), giving

$$\log_{10} N = \log_{10} N_0 - \lambda t \log_{10} e$$

or

$$\log_{10} N = -0.4343\,\lambda t + \log_{10} N_0.$$

This shows that if $\log_{10} N$ is plotted against t the result will be a straight line.

METHOD OF USING LOG GRAPH

To test whether N and t are related exponentially, add to your table of data a third row, of $\log_{10} N$, and plot this against t; if this log graph is a straight line you have an exponential relation. The y-intercept or data zero is $\log_{10} N_0$, so the value of N_0 can be found at once by taking its antilog. The gradient of the lines is $-0.4343\,\lambda$ from which the decay constant λ is easily calculated. Figure 10.3 displays the data of example 1 in this way. A similar result holds for exponential growth, but the gradient of the straight line is then positive. There is special log graph paper for this purpose, which saves the bother of looking up log tables. It is convenient to use but in no way essential.

Fig. 10.3.

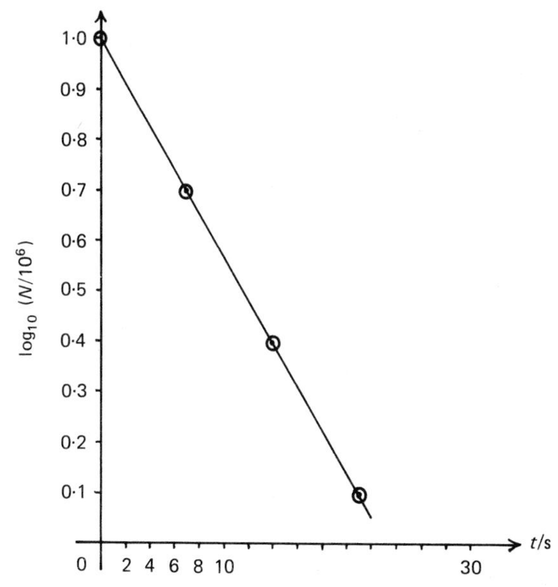

DIFFERENTIATION AND INTEGRATION OF e^x

We have seen that

$$\frac{dN}{dt} = -N_0 \lambda e^{-\lambda t}.$$

We state the general result that when $y = Ae^{kx}$, where k and A are constants,

$$\frac{dy}{dx} = Ake^{kx}, \quad \text{or} \quad \frac{dy}{dx}\Big/ y = k.$$

Since integration is the inverse operation to differentiation we see that

$$\int Ae^{kx}\,dx = (A/k)e^{kx} + \text{constant}.$$

Exercise B

1. A sample of a radioactive nuclide gives an average count-rate of $710\,s^{-1}$ when first placed in a counting device, and a count-rate of $120\,s^{-1}$ after 300 s. Calculate λ and $T_{1/2}$ for the nuclide.
2. The β^- activity from a sample of carbon prepared from a recently cut tree was recorded as $0.190\,s^{-1}$, whereas a similar sample from an old church gave a count-rate of $0.160\,s^{-1}$. Estimate the age of the church to the nearest 50 years, given that for ^{14}C, $T_{1/2} = 5580$ years.
3. Calculate the decay constant for element X given that $N = 0.310\,N_0$ when $t = 359$ s.
4. The rate of a chemical reaction is known to be $k = A\exp(-E/RT)$, where A, E and R are constants. What would be the most useful graph to plot to find E, given R?

Exercise C

Test the following "populations" for exponential variation and where relevant find λ and $T_{1/2}$.

1. In an experiment to measure the half-life of Radon-220, the count rate of this alpha emitting gas was measured.

Time/s	5	35	65	95	125	155	185
Counts per second/s^{-1}	56.2	44.1	32.7	23.2	16.8	13.3	9.6

2. (a) The following table gives the total public expenditure from 1951 to 1971, in the United Kingdom. (From the *Annual Abstract of Statistics*, Central Statistical Office, 1970, H.M.S.O.)

Year	1951	1961	1966	1968	1969	1970	1971
Expenditure in million £	5819	10307	15289	19092	19789	21595	24240

(b) The public expenditure on education is given in the following table.

Year	1951	1961	1966	1968	1969	1970	1971
Expenditure on Education in million £	398	1012	1768	2182	2347	2611	3006

In the following problems about the absorption of radioactive particles by matter, test the "populations" for exponential variation and when this occurs find the thickness of material for which the original count-rate is halved. This is commonly called the "half thickness".

3. In an experiment to investigate the absorption of γ rays by matter, lead absorbers of varying thickness were placed between the γ source and the radioactivity counter and the count rate measured.

Lead thickness/cm	0	0.32	0.635	0.95	1.27	1.59	1.90	2.22	3.18
Count rate/min^{-1}	464	392	328	263	229	183	166	126	78

4. In an experiment to investigate the absorption of beta particles by matter, aluminium absorbers of varying thickness were placed between the beta source and the radioactivity counter and the count rate measured.

Absorber thickness/mm	0	0.0125	0.0254	0.0381	0.0559	0.102
Count rate/min^{-1}	9421	8497	7339	6705	5599	2940

Absorber thickness/mm	0.163	0.203	0.254	0.28	0.343
Count rate/min^{-1}	1179	517	107	47	10

11

STATISTICS II

It may be tedious to calculate the mean and standard deviation of a population but the sense of producing order out of chaos is pleasant; an amorphous group of experimental data has been given shape and meaning. Much remains to be done, however, one big problem (only partially answered in chapter 2) being the estimation of experimental errors. Questions come to mind. How much reliance can be placed on the results of tests? Are two characteristics possessed by individuals (eye colour and intelligence, height and weight, month of birth and artistic ability) correlated or not? Did a certain fertiliser increase the yield of wheat significantly? What deductions can we make about a whole population from a small sample and how reliable are they? Is the difference between samples significant or not?

We hope to enable you to use statistics to obtain sensible answers to such questions, but of course the methods available are only tools. You have to formulate the question first and if it is the wrong question no amount of calculation will yield a meaningful answer. Sometimes the result seems obvious—height and weight are very likely to be strongly correlated for instance—and the tedious calculation necessary to produce a statistic only underlines this. So why waste time? Well, sometimes a simple test suffices; but sometimes you need to compare your work with that of others in the same field and to express in numerical terms your confidence in your result.

There is another sound reason for using statistical methods to analyse results rather than relying on simple examination of them. A test used in the right context carries much more weight than the experimenter's word alone; human optimism and natural bias are very hard to eliminate, familiarity with the conditions of the experiment may blind one to sources of error, and so on. The test, properly applied, is objective.

In such a small space we can take only a quick look at a few techniques. Later on you may have to study Statistics in greater depth in order to design effective experiments (much basic work in the subject was done by biologists for this purpose). Meanwhile unfortunately you have to take on trust the formulae thrust at you; you must see that you observe any necessary restrictions on their use and carry out instructions faithfully.

CONFIDENCE LIMITS

A certain population consists of the weights of 150 mice; the mean weight turns out to be $\bar{x} = 29.6$ g, the standard deviation $s = 1.2$ g, and this information only (not the whole frequency distribution) is available to you. What prediction can you make about the weight of a mouse picked at random from that group of 150? It is known that at least 75% of the population lies within two standard deviations of the mean, that is, in the range $\bar{x} \pm 2s$, but for almost all distributions encountered in scientific work about 95% is within this range. In this particular case 95% of the weights w are expected to be in the range $27.2\,\text{g} \leqslant w \leqslant 32.0\,\text{g}$. Note the word *expected*; we are dealing with probabilities, and do not suppose that exactly 142.5 weights will lie within the range, any more than we expect exactly 500 heads when a coin is tossed 1000 times. Expressed in terms of probability, the probability that a mouse chosen at random will have a weight less than 32.0 g and more than 27.2 g is 0.95, and we call this a 95% confidence limit.

That particular distribution was known to us only by its two statistics, $\bar{x}$ and s, and we could apply only the general result for the range containing 95% of the weights. If we know more about the distribution, either because we have all the records and can draw a histogram, or from theoretical considerations, we can find 98%, 99%, or 99.9% **confidence limits** (probabilities 0.98, 0.99, 0.999 respectively), using whichever is appropriate in each particular case.

There are several well-known distributions: *Uniform* (when every outcome is equally likely, as in throwing a die), *Binomial*, *Poisson* and *Normal*, are some whose properties are well established and whose parameters can be calculated theoretically. Mathematical models can be constructed to describe particular situations and then used to predict results. This is not the place for detailed discussion of such models, but for one, the Normal distribution, you do need more information.

Fig. 11.1.

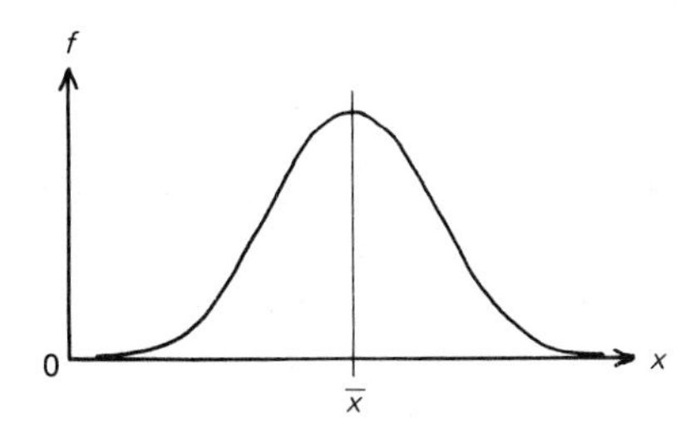

NORMAL DISTRIBUTION

It has been known for a long time that in certain circumstances the distribution of a variate takes a certain recognisable form, the histogram being symmetrical (figure 11.1). At various times the distribution has had different names, but is now almost always called *Normal.* It occurs whenever a large number of different factors are operating to cause the variations observed in the quantity being measured. A professional golfer driving a large number of balls from one tee on one afternoon would expect to achieve a certain length each time; however there are bound to be variations; wind changes, the backswing a fraction greater on one shot, the ball not hit quite along the intended line in another, the tee slightly higher, and so on. The result would show a distribution of lengths, approximately **normally distributed about the mean.** Similar influences give a **normal**

distribution of marks when an examination is taken by a whole year group in a school (or in the results obtained in an external examination). One pupil is inaccurate in a calculation, another misunderstands a question, a third draws a graph rather badly, yet all are about the same standard, and so the tally of small influences mounts up. Those mice would probably have weights normally distributed; continuous variation in phenotype (characteristics) results from the combination of many genetic and environmental factors.

The mathematical model is a continuous distribution whose mean is 0 and standard deviation 1, given by the equation

$$\phi(x) = (1/\sqrt{2\pi})\exp[-(\tfrac{1}{2})x^2].$$

Actual values of $\phi(x)$ are less important than values of areas under the curve, which represent probabilities: they are given in Normal Probability tables.

Fig. 11.2.

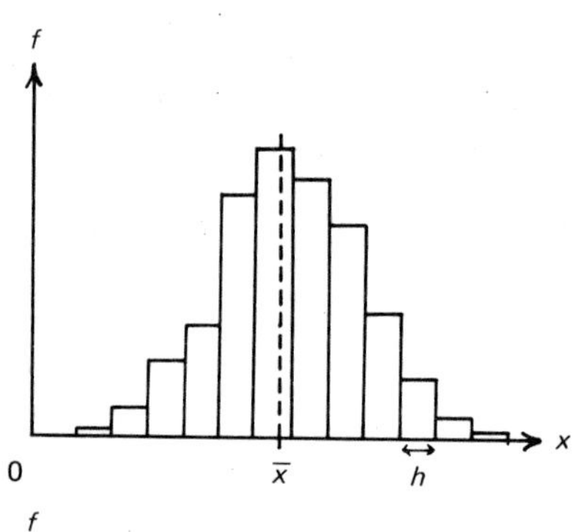

Fig. 11.3.

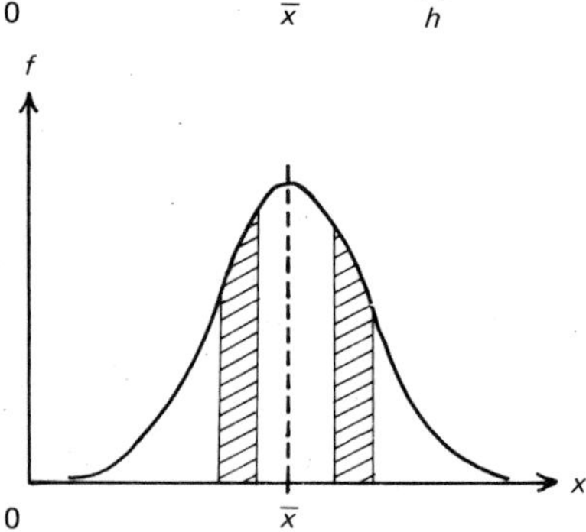

TABLE OF NORMAL PROBABILITY

To understand this table look first at the histogram in figure 11.2. If the population is large and the class intervals, h, small, then in the limit when $h \to 0$ the boundary of the histogram becomes a smooth curve (figure 11.3). The columns of figure 11.2 are replaced by strips under the curve (two are shaded in figure 11.3). Figure 11.4 shows the curve $y = \phi(x)$, the **normal probability curve,** which has the property that the whole area under it is 1. The mean value of x is 0, and since the curve is symmetrical about the mean the shaded area is 0.500; this means that the probability that $x < 0$ is 0.500. In figure 11.5 the shaded area (value 0.841) represents the probability that

Fig. 11.4.

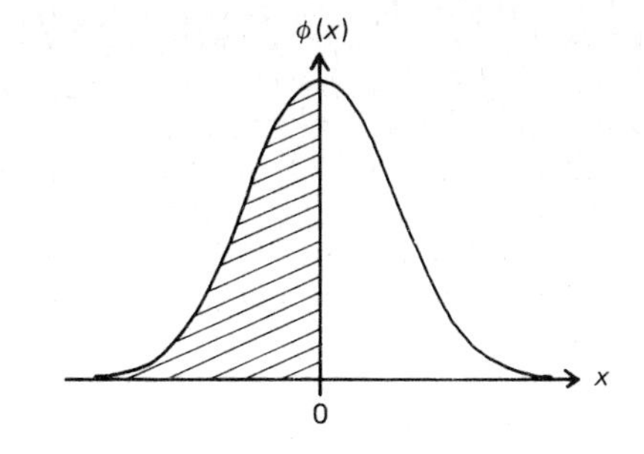

Fig. 11.5.

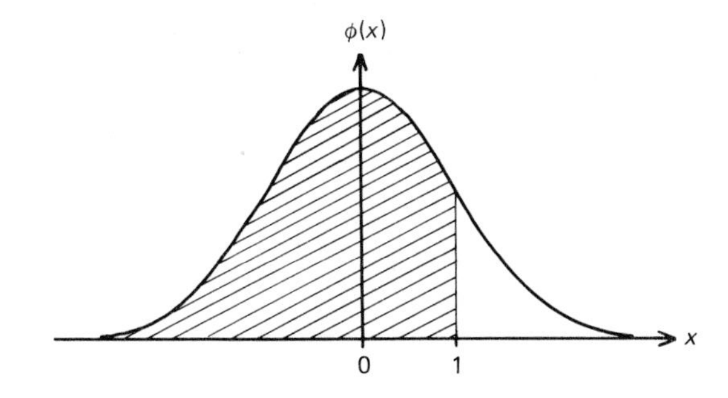

$x < 1$. Using the integral

$$\Phi(x) = \int_{-\infty}^{x} (1/\sqrt{2\pi})\exp\left(-(\tfrac{1}{2})y^2\right)dy, \quad x \geqslant 0$$

the values of these areas have been calculated and are recorded in the Normal Probability table.

HOW TO USE A TABLE OF NORMAL PROBABILITY

You must first be reasonably sure that your variate is normally distributed; then you need the mean, $\bar{x}$, and standard deviation, s, of your distribution. You may already have wondered about the utility of a distribution with mean 0 and standard deviation 1; but we *standardise* our variate x_i, giving

$$X_i = \frac{x_i - \bar{x}}{s}.$$

X_i expresses the deviation of x_i from the mean $\bar{x}$ as a number of standard deviations. We then use the table, taking the value for X_i. Example 1 should make this clear.

Example 1

Resistors of nominal value 100 ohms are found to have mean resistance $\bar{x} = 99.5$ ohms, standard deviation $s = 1.1$ ohm. Assuming that the distribution is normal, find the probability that a resistor chosen at random will (a) have resistance less than 101 ohms, (b) have resistance in the interval between 101.5 and 102 ohms.

(a) Standardising the variate for 101 ohms gives

$$X_1 = \frac{101 - 99.5}{1.1} = \frac{1.5}{1.1} = 1.36,$$

that is, 101 ohms is 1.36 standard deviations greater than the mean. Consulting the table shows that the required probability, $\Phi(1.36) = 0.913$. See figure 11.6(a).

(b) For 101.5,

$$X_2 = \frac{101.5 - 99.5}{1.1} = \frac{2.0}{1.1} = 1.82;$$

for 102

$$X_3 = \frac{102 - 99.5}{1.1} = \frac{2.5}{1.1} = 2.27.$$

From the table $\Phi(2.27) = 0.988$, $\Phi(1.82) = 0.966$, and the required

$$\begin{aligned} \text{probability} &= \Phi(2.27) - \Phi(1.82) \\ &= 0.988 - 0.966 \\ &= 0.022. \text{ See figure 11.6(b).} \end{aligned}$$

Fig. 11.6 (*a*)

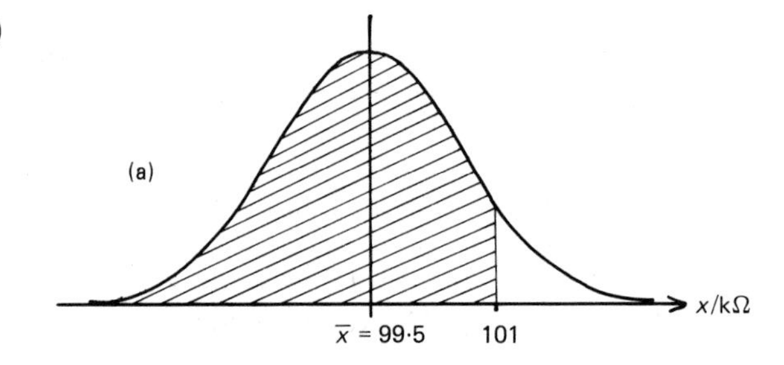

Fig. 11.6 (*b*)

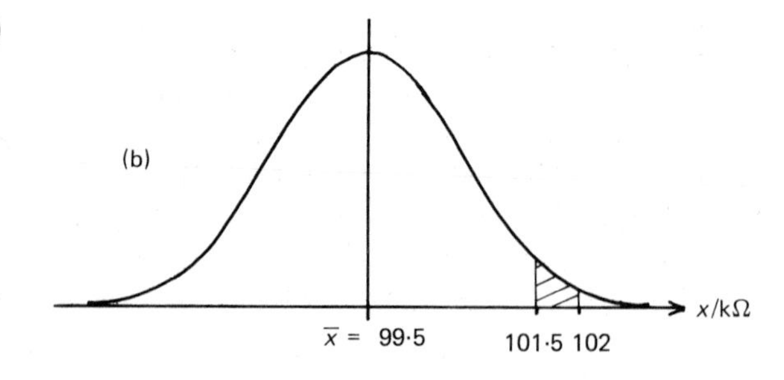

Fig. 11.7.

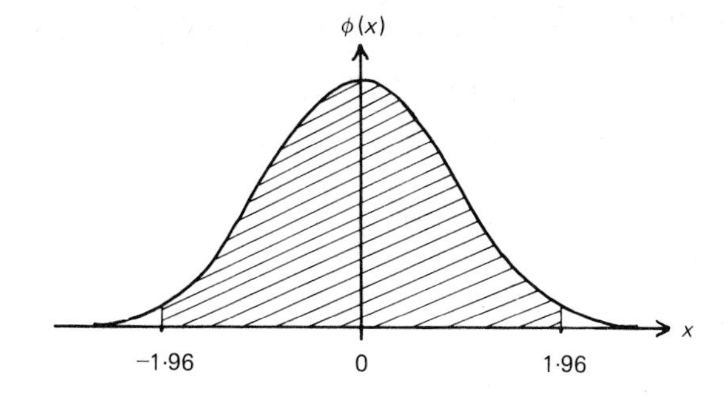

CONFIDENCE LIMITS FOR THE NORMAL PROBABILITY FUNCTION

95% CONFIDENCE LIMIT

The familiar shape of the normal probability function appears again in figure 11.7, with mean 0 and standard deviation 1. 95% of the area under the curve is shaded, the boundaries being at ± 1.96; each unshaded tail contains 2.5% of the total area. To check the boundaries find the probability

$$1 - 0.025 = 0.975; \quad 0.975 = \Phi(1.96).$$

This means that for a normal distribution mean $\bar{x}$, standard deviation s, 95% of the population lies in the range $\bar{x} \pm 1.96\,s$.

99% CONFIDENCE LIMIT

This time the unshaded area must consist of 1% of the area, each part being 0.5%. The problem is to find the value of x for which $\Phi(x) = 0.995$. From the table we find that $\Phi(2.58) = 0.9951$, and the interval for 1% confidence limit is $\bar{x} \pm 2.58\,s$. The 0.1% confidence interval similarly is $\bar{x} \pm 3.29\,s$, since $\Phi(3.29) = 0.9995$.

The following worked example illustrates some of the problems encountered.

Example 2

Intelligence quotient, IQ, is taken to have mean 100 and to be normally distributed. Over a certain population of students, the standard deviation is 15. What percentage of this population has IQ (a) over 120, (b) under 70, (c) between 130 and 140? (d) What is the probability that an individual, chosen randomly, has an IQ between 95 and 100? (e) What are the maximum and minimum IQs in the 95% confidence interval?

(a) IQ is quoted as a whole number, so in a histogram the centres of the columns will be at 98, 99,... and their ends at 97.5, 98.5, 99.5,.... Thus for IQ over 120 we standardise 120.5, and find

$$X_1 = \frac{120.5 - 100}{15} = \frac{20.5}{15} = 1.37.$$

$\Phi(1.37) = 0.915$, and percentage of the population with IQ > 120 is $100 - 91.5 = 8.5\%$.
(b) For IQ under 70 we must standardise 69.5:

$$X_2 = \frac{69.5 - 100}{15} = \frac{-30.5}{15} = -2.03.$$

$$\begin{aligned} \Phi(2.03) &= 0.979 \\ \Phi(-2.03) &= 1 - \Phi(2.03) \\ &= 1 - 0.979 \\ &= 0.021, \end{aligned}$$

and the percentage of the population having IQ < 70 is 2.1%.
(c) The variates to be standardised are 129.5 and 140.5.

$$X_3 = \frac{129.5 - 100}{15} = \frac{29.5}{15} = 1.97$$

$$X_4 = \frac{140.5 - 100}{15} = \frac{40.5}{15} = 2.70.$$

$$\begin{aligned} \text{Required probability} &= \Phi(2.70) - \Phi(1.97) \\ &= 0.9965 - 0.976 \\ &= 0.020 \end{aligned}$$

and the percentage of the population whose IQ lies between 130 and 140 is 2.0%.
(d) The variates are 94.5 and 100.5.

$$X_5 = \frac{94.5 - 100}{15} = -\frac{5.5}{15} = -0.37$$

$$X_6 = \frac{100.5 - 100}{15} = \frac{0.5}{15} = 0.03.$$

$$\begin{aligned} \Phi(-0.37) &= 1 - \Phi(0.37) = 1 - 0.644 \\ \Phi(0.03) &= 0.512 \\ \Phi(0.03) - \Phi(-0.37) &= 0.512 \;\; - (1 - 0.644) \\ &= 1.156 - 1 \\ &= 0.156, \end{aligned}$$

the required probability.
(e) 95% of the population will have IQs within the range $100 \pm 1.96\,s$, namely $100 \pm 1.96 \times 15$.

$$1.96 \times 15 = 29.40.$$

The range is therefore $100 - 29.40 \leqslant \text{IQ} \leqslant 100 + 29.40$, or $71 \leqslant \text{IQ} \leqslant 129$.

Example 3
Assuming that our population consists of 200 mice whose weights are normally distributed about the mean, 29.6 g, with standard deviation 1.2 g, find the number of mice which we would expect to have weights of less than 28 g.

Weight is a continuously varying quantity. We have no information about class intervals or the accuracy with which the measurements were done, so we simply take the variate 28 and standardise it:

$$X = \frac{28 - 29.6}{1.2} = -\frac{1.6}{1.2} = -1.333.$$

$\Phi(1.333) = 0.908$, $1 - 0.908 = 0.092$, and the expected number of mice with weights below 28 g is $0.092 \times 200 = 18.4$, namely about 18.

In practice many distributions are found to agree well with the normal over a wide domain of X. Nevertheless many concern discrete variates and all are finite. If you look back at the expression for $\Phi(x)$ you will find that the lower limit of the integral is $-\infty$, and from $\Phi(x)$ we know that there is no finite upper limit either; consequently no finite distribution is normal. We are therefore adopting a mathematical model, well tried, and tacitly restricting the domain of X in which we use it to a region where the fit (theoretically and practically) is good.

In example 2 there is a finite lower limit, the lowest measurable IQ. We do not know the upper limit or even whether there is one; Sci Fi writers have explored the consequences of individuals having very high IQs (an IQ of 250 is already 10 standard deviations from the mean, probability very small but not zero), but in an actual investigation with a finite population there will be an upper limit.

The mouse weights of example 3 also have a finite lower limit, 0 g. That odd creature the reasonable man would accept that a finite upper limit exists even if it is not known, so here is a finite distribution again, agreeing very well with the normal over a wide range of values of X.

The distinction between finite and infinite distributions is seldom important in practice because of the rapid falling off of the tail (the probability of $|x| > 4$ is only 0.00006), but if you are concerned with large deviations from the mean you must give very careful thought to the nature of the distribution for such X.

Exercise A

1. For each of the following distributions standardise the variates given.
   (a) mean 7.00, s 0.3; standardise 6.00, 7.20, 7.42,
   (b) mean 150, s 25; standardise 170, 110, 205,
   (c) mean 2300, s 112; standardise 2400, 2600, 2000.
2. Establish the 95% confidence interval for a normally distributed variate whose mean is 16 and standard deviation 4.
3. Butter is sold in 250 g packets, labelled minimum net weight 250 g. The packaging machine is set to 254 g and tests show that the standard deviation of the weights of the packets is 2.2 g. What proportion of the butter packets will be rejected as being less than 250 g?

CORRELATION

You notice two characteristics of some seedlings of wheat, height and age, or you watch children playing and wonder whether the noisy ones are also the most energetic, and in each case you link the two qualities in your mind, wondering whether they are correlated. To give a valid answer to your speculation about the children would be very difficult since the two characteristics involved are not readily measured. But for the seedlings you could measure height h and age T for each one and arrive at a statistic describing the degree of correlation between h and T.

The question "are h and T correlated for this population of wheat seedlings?" is a sensible one, and can be investigated numerically since both quantities can be measured for each individual. The other question, "are noisy children energetic?" is prompted by observation and quite sensible, but without considerable resources you cannot go further. There is a third type of question; for instance, "in one large city is there correlation between the fish consumption per annum and the number of 'A' level passes gained by sixth form pupils?" Both characteristics have numerical values so that a statistic could be worked out and might be significant without necessarily being meaningful. As always statistics can give an answer, but it is meaningless if the original question was foolish, biased, or the scientific method faulty.

SCATTER DIAGRAM AND CORRELATION COEFFICIENT

It is often enough to plot a **scatter diagram** to find out whether two quantities are correlated. In the case of the wheat seedlings plot one point (h, T) for each individual. If there is correlation the points will lie near a straight line; if not, they will be randomly scattered. Three types of graph occur; 11.8(a) shows positive correlation, in 11.8(b) there is no correlation and 11.8(c) shows negative correlation.

In many cases a qualitative demonstration of correlation is all that is needed, and the scatter diagram is sufficient. However you sometimes need to compare your work with other people's results, or the scatter diagram does not obviously indicate correlation. Then you must use the **correlation coefficient r**, defined by

$$r = \frac{\sum_{i=1}^{n} (x_i - \bar{x})(y_i - \bar{y})}{\sqrt{\sum_{i=1}^{n} (x_i - \bar{x})^2 \sum_{i=1}^{n} (y_i - \bar{y})^2}}. \qquad (1)$$

The population consists of n individuals each of which has two variates x_i and y_i whose means are $\bar{x}$ and $\bar{y}$. In the denominator, always take the positive square root; the numerator may be positive, negative or zero, and we find that $-1 \leqslant r \leqslant 1$. The extreme cases are listed below:

$\mathbf{r} = +1$. x and y are linearly related and all the points (x_i, y_i) lie on a straight line with positive gradient.

$\mathbf{r} = 0$. There is no correlation.

$\mathbf{r} = -1$. x and y are again linearly related, but this time the points (x_i, y_i) all lie on a straight line with negative gradient.

At the experimental stage do remember that if you want to test for correlation you must know which x and which y belong to each individual; it is no use having the x distribution and the y distribution without

Fig. 11.8.

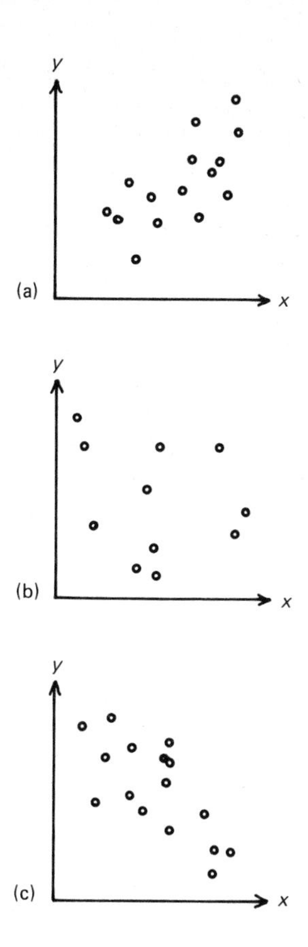

knowing which belongs to which. It is always advisable, too, to draw a scatter diagram before embarking on the calculation of r. If it resembles 11.8(b) r will be very small and it is probably not worth the tedium of the calculation.

The method for calculating r is shown in a worked example (example 5) but you might feel more willing to study this if you knew how to use r before beginning. So example 4 shows how to use the table once you know r.

USING THE TABLE FOR THE CORRELATION COEFFICIENT r (APPENDIX C)

$-1 \leqslant r \leqslant 1$, but in using the table for r only the numerical value is needed; the table gives a measure of the strength of correlation between the variates or the level of confidence with which one may say that there is a linear relation between them. Take a look at the table for r: the columns headed by probabilities p ranging from $p = 0.1$ to $p = 0.001$ contain values of r, with the appropriate degree of freedom (explanation later) given on the left. Find 7 in the left hand column (a small population); for $p = 0.01$, $r = 0.798$.

If you had 70 degrees of freedom, for the same probability $r = 0.302$ which confirms the common sense judgement that the smaller the population the greater $|r|$ must be to establish significance of correlation.

DEGREES OF FREEDOM

You may well ask why the left-hand column is not headed *number in population* or merely n; to give a partial answer look back to equation (1). In calculating $\bar{x}$ every one of the n variates x_i has equal weight, but when it comes to the deviations $x_i - \bar{x}$, once you have formed $n - 1$ of them the last is known (it can be calculated from the first $n - 1$), so you really now have only $n - 1$ independent quantities; or using the new term, **n − 1 degrees of freedom**. Thus in calculating r there are $n - 1$ degrees of freedom (the Greek letter v pronounced nu, is often used for number of degrees of freedom). So why not use the heading $n - 1$? You could perhaps, but in some applications processing of data reduces the number of degrees of freedom more drastically, and it is better to use the term and to remember that in this case it is $n - 1$.

For small populations the distinction between n and $n - 1$ is important (roughly for $n \leqslant 20$) and the table has separate rows for each value. As n increases it becomes merely pedantic to insist on the difference between n and $n - 1$ degrees of freedom (between 71 and 70 for instance) and rows are printed first at intervals of 5, then of 10 (more extensive tables exist of course).

Example 4

Test the significance of r in the following cases

(a) Population $n = 19$ $r = 0.400$;

(b) Population $n = 50$ $r = -0.400$;

(c) Population $n = 6$ $r = 0.820$.

(a) Number of degrees of freedom $= 19 - 1 = 18$. In the row for 18, 0.400 lies between the entries for 0.1 and 0.05 probabilities so the significance level for $r = 0.400$ lies between 10% and 5%, and this does not suggest a significant degree of correlation.

(b) Number of degrees of freedom $= 50 - 1 = 49$: in the row for 50 the entry for probability $= 0.01$ is 0.354 and for 0.001 it is 0.443. The numerical value of r, 0.400, lies between these and is significant at the 1% level and we can assume significant correlation. (A similar result is obtained from the 45 row, showing that there is no need for great precision in degrees of freedom when n is large).

(c) Number of degrees of freedom $= 6 - 1 = 5$. For $r = 0.820$ probability lies between 0.05 and 0.02, and r is significant at the 5% level only. In this case, had we used the 6 row, this value of r would have a probability between 0.02 and 0.07 and we might have placed too much reliance on this correlation.

Example 5

In a test given to a group of students, two different marking schemes A and B were used. The final positions for 11 students are given under both schemes. Is there significant correlation?

Pupil	a	b	c	d	e	f	g	h	i	j	k
Position A	1	2	2	2	5	6	7	7	7	10	11
Position B	1	2	2	5	2	6	7	8	10	10	8

It is obvious that the two schemes gave substantially the same ordering (not even a scatter diagram is needed to show this) and we expect a high value for r. With such a small population we should calculate r, however.

x	y	$x_i - \bar{x}$	$y_i - \bar{y}$	$(x_i - \bar{x}) \times (y_i - \bar{y})$	$(x_i - \bar{x})^2$	$(y_i - \bar{y})^2$
1	1	−4.45	−4.55	20.25	19.84	20.66
2	2	−3.45	−3.55	12.25	11.93	12.57
2	2	−3.45	−3.55	12.25	11.93	12.57
2	5	−3.45	−0.55	1.88	11.93	0.30
5	2	−0.45	−3.55	1.61	0.21	12.57
6	6	0.55	0.45	0.25	0.30	0.21
7	7	1.55	1.45	2.25	2.39	2.12
7	8	1.55	2.45	3.79	2.39	6.02
7	10	1.55	4.45	6.88	2.39	19.84
10	10	4.55	4.45	20.25	20.66	19.84
11	8	5.55	2.45	13.61	30.75	6.02
60	61			95.27	114.73	112.73

$$\bar{x} = \frac{60}{11} = 5.45, \qquad \bar{y} = \frac{61}{11} = 5.55,$$

$$r = \frac{95.27}{\sqrt{114.73 \times 112.73}} = 0.838.$$

Number of degrees of freedom $= 11 - 1 = 10$; $0.838 > 0.823$, the entry under $p = 0.001$ and there is significance at the 0.1% level.

Either scheme proves to be satisfactory and some other criterion is needed to decide which to use.

Exercise B

1. Calculated values of the correlation coefficient, r, together with the number of observations, N, are given for 5 experiments. For each, first write down v and then use tables for r to state whether you think there is significant correlation. Give the confidence limits. (a) $r = 0.86$, $N = 5$; (b) $r = -0.50$, $N = 20$; (c) $r = 0.50$, $N = 11$; (d) $r = 0.35$, $N = 30$; (e) $r = 0.24$, $N = 90$.

   For all the data of questions 2 and 3 plot scatter diagrams, and for *at least* one, calculate r, using tables to enable you to state whether there is a significant degree of correlation (positive or negative) taking as significance level 5%.
2. The table shows the height and middle finger length (cm) of each of 15 girls.

Height/cm x	165	161	169	154	160	164	154	161
Finger length/cm y	11.7	12.3	12.4	12.1	11.1	11.5	10.3	11.2

Height/cm x	154	167	169	170	169	172	167
Finger length/cm y	9.7	12.2	12.5	12.7	13.3	12.5	11.5

3. The relative distribution of plant species on dune slacks was measured, using 10 cm × 10 cm quadrats randomly placed, by determining the percentage cover of each species within the quadrat. There were 48 quadrats in each of a series of slacks, starting near the beach and working inland. The results are summarised in the table given below, and the problem is to decide whether the frequency of any of the plants is correlated with a high pH value.

Percentage cover (mean for 48 *quadrats)*

Dune slack	1	2	3	4	5	6	7	8
(a) **Galium verum** (lady's bedstraw)	3.0	3.0	11.5	6.6	0.9	0.3	6.8	6.6
(b) **Anthyllis vulneria** (lady's fingers)	3.4	0.3	1.6	4.6	1.5	0.5	0	0
(c) **Trifolium repens** (white clover)	0	0	0	0	2.2	0.7	0.6	0
(d) Mosses	0.7	4.2	8.0	2.4	0	1.5	0.1	3.0
pH	8.5	8.5	9.5	8.5	6.5	7.5	8.5	9.0

SAMPLING AND SAMPLE MEANS

Sampling and samples are familiar words and for once common usage and technical meaning are similar. A cook tasting the broth is taking a sample and using it to estimate the flavour of the whole. The free sample of a drug in pill form is supposed to enable the doctor to judge the efficacy of the whole population of pills from the sample of 20 or so. There is one definition we should mention: in a **simple sample** each element is taken out and recorded, then replaced before the next is drawn. Neither of our examples could be a simple sample and in practice most are not but sampling without replacement is difficult to deal with theoretically. Fortunately when the number in the population is large compared with the sample size all samples may be treated as simple samples. How to take samples will be dealt with briefly under the heading *Sampling Techniques*; in this section we confine ourselves to extracting information from them.

We consider a population whose mean (μ) and standard deviation (σ) we cannot calculate, and take as an example the long diameters of the eggs of herring gulls. It is inconceivable that we could find, let alone measure, all the eggs of all existing herring gulls, so we measure a number of samples of n eggs and estimate μ and σ from the samples. The samples have means m_1, m_2, m_3...and the average of these means is $\bar{m}$ the mean of the sample means. This is the best estimate of μ which we can obtain, and even if we can only have one sample, its mean is still the best available estimate of μ. What we have obtained is a **distribution of sample means** with mean $\bar{m}$ and standard deviation σ_n, both of which can be calculated. The distribution of means tends to be normal even if the population as a whole is not (although we would expect the n elements of each individual sample to mirror more or less accurately the distribution of the parent population). From σ_n we can estimate σ, from the result $\sigma_n = \sigma/\sqrt{n}$.

Most often we have only one sample of n elements, mean m, standard deviation (S.D.) s. Then we must use m as our estimate of μ, but how reliable an estimate is it? m is just one value belonging to the distribution of means which, as we have seen, has standard deviation $\sigma/\sqrt{n}$. We can use $s/\sqrt{n}$ as an estimate of this, and since it gives us a measure of how reliable a value m is, $s/\sqrt{n}$ is called the Standard Error or S.E.

This idea was used in chapter 2 when it was mentioned that $\Delta x/\sqrt{n}$ is a typical or standard error of the mean of n results. Δx is now best identified with the standard deviation of individual results.

COMPARISON OF SAMPLE MEANS; STUDENT'S t

The need often arises to decide whether or not there is a significant difference between two samples A and B—are they taken from two different populations or from the same one? There is no easy way of doing this, (superficial judgements are often misleading), but a most reliable statistic (used correctly) is t, often called Student's t.

Two samples A and B, containing N_A and N_B elements have means m_A and m_B whose difference is $m_A - m_B$. To find whether this is large enough to be significant we need to know the standard error of this difference, which we call s, where

$$s = \sqrt{\frac{s_A^2}{N_A} + \frac{s_B^2}{N_B}}.$$

Then we standardise the variate $m_A - m_B$, calling the standardised variate t, so that

$$t = \frac{m_A - m_B}{s}.$$

To see how this works, study the next example.

Example 6

The number of flowers on a bluebell inflorescence varies greatly. Is sunlight an important factor affecting this? Compare two samples, the first, A, containing 30 inflorescences from full sunlight, the second 31 inflorescences growing in partial shade, to find out.

No. of flowers on inflorescence	4	5	6	7	8	9	10	11	12	13	14
Frequency A	2	3	5	3	4	5	2	2	0	3	1
Frequency B	2	4	5	14	2	1	1	2	0	0	0

First calculate means and variances for A and B.

$$N_A = 30 \qquad N_B = 31;$$
$$m_A = 8.200 \qquad m_B = 6.871;$$

$$s_A^2 = 7.683 \qquad s_B^2 = 2.849;$$
$$s_A = 2.772 \qquad s_B = 1.688.$$

In finding the variances the formula used was

$$s^2 = \frac{\sum_{i=1}^{n} f_i(x_i - \bar{x})^2}{N-1},$$

dividing by $N - 1$ instead of N.

This is strictly correct because, as in calculating r, there are only $N - 1$ degrees of freedom. The difference between the result of dividing by N and by $N - 1$ is usually far less than the accuracies involved, but now that you are familiar with the idea of degrees of freedom we have used that formula for s^2 which is theoretically correct. The two means are within one S.D. of each other but it is their difference and its S.E. which matter.

$$t = \frac{m_A - m_B}{s}; \qquad s = \sqrt{\frac{s_A^2}{N_A} + \frac{s_B^2}{N_B}}$$
$$= \sqrt{\frac{7.683}{30} + \frac{2.849}{31}}$$
$$= 0.590,$$

$$\Rightarrow t = \frac{8.200 - 6.871}{0.590}$$
$$= 2.25$$

(it does not matter whether t is positive or negative). The difference of the means is more than 2 standard errors; using the Normal probability tables,

$$\Phi(2.25) = 0.9878,$$
$$1 - 0.9878 = 0.0122,$$

and the samples may be said to differ significantly (at the 1.22% level).

For samples of $N < 30$ there are special tables of Student's t, or t-distribution, to be used in the same way as those for r. For these you need to know the degrees of freedom, $v = N_A - 1 + N_B - 1$. Even if the sample is small, a large value of t indicates a significant difference without resorting to tables; at $v = 10$, $t = 3$ is significant at 1% level. If you are in charge of sampling try to have samples of the same size. If $N_A = N_B = N$ the formula for s reduces to

$$s = \sqrt{\frac{s_A^2 + s_B^2}{N}}.$$

STUDENT'S t; SUMMARY

Use t to test whether two samples A and B (size N_A and N_B, means m_A and m_B, standard deviations s_A and s_B), come from the same or different populations. Calculate $s = \sqrt{(s_A^2/N_A.) + (s_B^2/N_B)}$. Then $t = (m_A - m_B)/s$. When N_A and N_B are each not less than 30 (approximately) consult Normal Probability tables to find the significance levels at which you can say that the samples differ.

For sample sizes less than 30 consult instead the special tables for the t-function. To do this you need the number of degrees of freedom. $v = N_A - 1 + N_B - 1$.

Exercise C

1. Two feeds, A and B, are tested, A being fed to 32 rats, B to 35 rats, the animals being kept in conditions otherwise closely similar. The gains in weight were recorded. For A it was found that the mean, m_A, was 52.0 g, standard deviation $s_A = 8.00$ g. In the case of feed B, $m_B = 58.1$ g, $s_B = 9.81$ g. Use the statistic t to compare the means of the two samples and hence decide whether they differ significantly (5% significance level).
2. Different methods of teaching are used in two schools, A and B. To enable a comparison to be made, the same test was given to two groups, one in each school, and the mean scores tested using Student's t (as far as possible the groups were chosen to be similar in composition).

   The results were as follows:

	Number in Group	Mean	Standard Deviation
A	42	52.3	18.1
B	43	65.2	19.0

Apply Student's t to these results and state whether the test shows that the methods used in school B are producing significantly better results than those in A.

TESTING A HYPOTHESIS—THE CHI-SQUARED TEST

When two coins are tossed theory predicts that the numbers of times that 2 heads, 1 head and 0 heads will appear should be in the ratio 1:2:1. An actual experiment yields the numbers 9, 32 and 15 in these three categories. How can we judge whether this result is consistent with the expectation? First we determine the total number of observations, 56, and divide this number in the expected ratio 1:2:1, giving 14, 28 and 14 in the three categories; we now have 14 to compare with 9, 28 with 32 and 14 with 15. χ^2, described below, enables you to do this.

The statistic χ^2 (Greek letter chi, pronounced ky) tests experimental results divided into categories against a hypothesis which predicts the numbers in each category. To calculate χ^2, take the deviation of each observed value, O, from its expected value, E, square this deviation and divide it by the expected value; then sum over the categories. The formal definition is thus

$$\chi^2 = \sum\{(O - E)^2/E\}.$$

There are tables for χ^2 which are used in a way somewhat similar to those for r, but—and this is important—there are two conditions which must be satisfied before their use is justified. These are, first, that not less than 50 observations are available, and second, that each category must contain at least 10 observations (but categories containing less than 10 can be grouped together to form a more inclusive one). Both 50 and 10 are approximate; one might use the test with 48 observations for instance, but not with 35. In example 7 we calculate χ^2 for the experiment we have already described to illustrate the method.

Example 7
Test the hypothesis that the experimental results 9, 32, 15 are not significantly different from those expected in each category, namely 14, 28, 14. Do this by calculating χ^2 (first justifying its use) and using the tables to help you to form your judgment.

Find first the total number of observations, N (56), then the lowest number in any category (9). $N > 50$, and 9 is nearly 10 so we may use the χ^2 test.

We already know the expected values, E, so go on to find the number of degrees of freedom, v. When the first 2 categories have been filled (here 14 and 28) the third is determined by the total which must be 56, so $v = 2$. Generally, for n categories, $v = n - 1$.

Now we complete the calculation, noting that the totals for O and E are both 56 and for $(O - E)$ the total must be zero. There is no need for separate rows for $(O - E)^2$ and $(O - E)^2/E$; they are given here for completeness.

		Categories		
	2 heads	*1 head*	*0 heads*	*total*
O	9	12	15	56
E	14	28	14	56
$O - E$	−5	4	1	0
$(O - E)^2$	25	16	1	
$(O - E)^2/E$	25/14	16/28	1/14	68/28 = 2.43

$\chi^2 = 2.43$. Consult the table in the row for $v = 2$. For probability $p = 0.50$ we find that $\chi^2 = 1.386$, and for $p = 0.20$, $\chi^2 = 3.22$. Our value, $\chi^2 = 2.43$, lies between these, so no doubt is cast on our hypothesis by this statistic. What the tables actually tell us, and why this is so, requires a little explanation. Assume that the hypothesis is true and that the experiment can be repeated a number of times. Then in 50% of these replications a value of $\chi^2 > 1.386$ will occur, and in 20% of them we will find that $\chi^2 > 3.22$. Thus our value, 2.43, is quite consistent with the hypothesis that the ratio is 1:2:1.

TESTING A NULL HYPOTHESIS

In example 7 we did not expect to upset our hypothesis; sometimes we adopt an opposite attitude and test a **null hypothesis**; this predicts expected values on the assumption that no disturbing effects exist, and we hope to show that the hypothesis is inadequate. The object of any experiment could be rephrased in terms of a null hypothesis; for example, "an experiment to determine the effect of increased carbon dioxide concentration on the breathing rate of a locust" could be restated, using the null hypothesis, as "increased carbon dioxide concentrations have no effect on the breathing rate of a locust". In example 8 a die was thrown and the scores recorded; the results look suspiciously non-uniform. We adopt the null hypothesis that all the scores are equally likely (a perfect die) and if our suspicions of a biased die are justified χ^2 will be large because there will be a wide discrepancy between observed values and those predicted using the null hypothesis.

Example 8

The table below gives the scores obtained in 132 throws of a die. Take as null hypothesis that the die is perfect, calculate χ^2, and use the χ^2 table to express your confidence or otherwise in the die.

	Score						
	1	2	3	4	5	6	*Total*
0	25	16	26	12	20	33	132
E	22	22	22	22	22	22	132
$(0 - E)$	3	−6	4	−10	−2	11	0
$(0 - E)^2$	9	36	16	100	4	121	
$(0 - E)^2/E$	9/22	36/22	16/22	100/22	4/22	121/22	286/22 = 13.0

To avoid repetition the whole table of calculations is given together above. In practice you have as data only the *0* row and begin by justifying the use of the test; $N = 132$ and the lowest number in any category is 12 so you can go ahead to calculate the expected values E. In this case they are all $132/6 = 22$, and $v = 6 - 1 = 5$. The value arrived at for χ^2 is 13.0.

Consulting the tables we find, for $v = 5$, that $p = 0.05$ gives $\chi^2 = 11.07$ and $p = 0.02$ gives $\chi^2 = 13.39$. This shows that our null hypothesis will produce $\chi^2 = 13.0$ in only about 2% of experiments on perfect dice. There is no proof that the die we used was biased, but it would be sensible to reject it on the basis of this test.

LOW VALUES OF CHI-SQUARED

We saw in example 7 that a small value for χ^2 indicated that observed results were consistent with the expected values. However, a very low value of χ^2 should not be accepted uncritically as extra-convincing confirmation of the hypothesis; instead it should send you back to your experiment for a long, hard look at your scientific method. The next example illustrates why this is so.

Example 9
A student submits data, collected at home, in an experiment similar to that in example 8, but for 120 throws. Test the hypothesis that the die is a fair one, and comment (the actual calculation of χ^2 is left for you to do for yourself).

Score	1	2	3	4	5	6
Observed frequency	21	18	19	19	21	22

$N = 120$, the lowest number in any category is 18 and $\nu = 5$. $\chi^2 = 0.600$; for $\nu = 5$ and $p = 0.99$, $\chi^2 = 0.554$. This value, $p \approx 0.99$, is not impossible but it is very high indeed, and as the experiment was not performed under controlled conditions it must be rejected; the hypothesis that the student simply wrote down a likely set of numbers is emphatically not disproved; the lecturer will know better than to set such an experiment for private study in the future.

CONTINGENCY TABLES

So far we have used χ^2 to test the results, recorded in categories, of one experiment against those predicted by a chosen hypothesis. There are many cases where two or more different sets of numbers in the categories make up the results of an experiment so that the table has two or more rows; the name given to such a table is a **contingency table**. In an extension of the experiment described in example 7 for instance, you might try spinning the two coins on a table—first row—and throwing them up, allowing them to fall on the ground—second row. The new table shown below has two rows and three columns and is a **2 × 3 contingency table.**

	2 heads	*1 head*	*0 heads*	*total*
Spun	9	32	15	56
Tossed	16	25	9	50
Total	25	57	24	106

In the previous examples all the tables have been $1 \times n$ and although there is no essential difference between them and those with more than 1 row the name contingency table is only given to the latter type.

The coin results were obtained in an experiment and provide a perfectly good example of a 2×3 contingency table, but there is not much interest in testing whether the two methods of tossing give consistent results; you can see at a glance that the second row is comparable with the first which we have already tested. In examples 10 and 11 we give full details of the calculation of χ^2 for two very different contingency tables. Both are 2×2, but this is a common type and demonstrates the method satisfactorily.

Example 10

The table below relates to an epidemic of smallpox (imaginary). Does this evidence confirm that vaccination is effective, as medical knowledge suggests? Using a suitable null hypothesis calculate χ^2 to enable you to decide.

Observed (*O*) Table

	Recoveries	*Deaths*	*Total*
Vaccinated	1715	162	1877
Non-vaccinated	79	65	144
Total	1794	227	2021

We first decide whether or not we can use the chi-squared test. $N > 50$ for the non-vaccinated group and the smallest number in any category is 65, so we can go ahead (in a contingency table with 2 or more rows you cannot use this test at all if even 1 entry is too small). We need to know v, but its value will appear when we construct the Expected (E) table.

We adopt as null hypothesis that vaccination has no effect. The *Total* row of the table shows that of the 2021 affected, 1794 recovered and 227 died, so we assume that the fraction of any group which will recover is 1794/2021. The full Expected (E) table is shown below in the form of fractions for each group, but you do not need to record the E table in this form yourself.

Expected (E) Table

	Recoveries	*Deaths*	*Total*
Vaccinated	$1877 \times (1794/2021)$	$1877 \times (227/2021)$	1877
Non-vaccinated	$144 \times (1794/2021)$	$144 \times (227/2021)$	144
Total	1794	227	2021

The totals impose restrictions on the tables since in the E table all the totals must be the same as those in the *O* table; consequently you need to calculate only one of the four fractions, obtaining the other three entries by subtraction. Thus there is only one degree of freedom ($v = 1$), and the calculation of the E table follows.

$1877 \times (1794/2021) = 1666$, but we could have chosen any one of the four to calculate.

Expected (E) Table

	Recoveries	*Deaths*	*Total*
Vaccinated	1666	$1877 - 1666 = 211$	1877
Non-vaccinated	$1794 - 1666 = 128$	$144 - 128 = 16$	144
Total	1794	227	2021

Now find $(O - E)$. The whole table is given to remind you that

$$\sum (O - E) = 0;$$

you will see at once that you need do only one calculation, but it is as well to perform all four subtractions as a check.

$(O - E)$ Table

	Recoveries	*Deaths*	*Total*
Vaccinated	49	−49	0
Non-vaccinated	−49	49	0
Total	0	0	0

$(O - E)^2 = 2401$, and combining this with the E table results we have

$$\chi^2 = \sum (O - E)^2/E = 2401(1/1666 + 1/211 + 1/128 + 1/16)$$
$$= 181.$$

There is a very small chance indeed that this value of χ^2 would arise if our null hypothesis were true. For $\nu = 1$ even at $p = 0.001$, $\chi^2 = 10.83$ only.

Example 11
Two plant species, A (sheep's fescue) and B (perennial oat-grass) grow in a field grazed by cattle, where 200 quadrats have been taken at random. Are A and B found in association? For each quadrat the records show A present alone, A and B present, B present alone, neither A nor B present, and it appears that A is recorded in 50 quadrats, B in 80. Taking as null hypothesis that there is no association, use the chi-squared test (if possible) to establish association or not, stating the significance levels clearly.

Observed (O) Table

		(A) Sheep's fescue		
		Present	*Absent*	*Total*
(B)	Present	26	54	80
Oat-grass	Absent	24	96	120
	Total	50	150	200

There are 200 quadrats, 80 in the smaller group, and the smallest number in any category is 24, so we can continue. $\nu = 1$. If there is no association the presence or absence of B has no effect on A, and the E table follows.

Expected (E) Table

		(A) Sheep's fescue		
		Present	*Absent*	*Total*
(B) Oat-grass	Present	20	60	80
	Absent	30	90	120
	Total	50	150	200

$$(0-E) = \pm 6, \text{ and we have}$$
$$\chi^2 = \sum(0-E)^2/E = 36(1/20+1/60+1/30+1/90)$$
$$= 3.24.$$

We enter the table at $v = 1$ and see that for $p = 0.1$ $\chi^2 = 2.71$, and for $p = 0.05$, $\chi^2 = 3.84$. The results are not inconsistent with our null hypothesis, and further investigation would be needed.

SUMMARY; CHI-SQUARED TEST

Use this test when results are recorded in categories, and you want to find whether or not they are consistent with a hypothesis set up by you.

Restrictions: the total number of observations, N, must be at least 50 (approximately) and the smallest number in any category must be at least 10. In tables with only one row you can however combine categories, but in contingency tables you cannot do this. You may find lower limits given in some texts, but those given here allow a safe margin. Set up the hypothesis and use to calculate the whole of the Expected (E) table.

The number of degrees of freedom, v, is determined while constructing the Expected table; when there are n categories and only one row $v = n - 1$; for a contingency table $m \times n$

$$v = (m - 1) \times (n - 1), \quad m, n > 1.$$

$$\chi^2 = \sum(0-E)^2/E,$$

so for each category calculate successively $(0 - E)$, $(0 - E)^2$ and $(0 - E)^2/E$ and then sum.

The significance level should be decided upon next. Notice that having chosen 5% for example it would be foolish to reject a hypothesis on the basis of a result of 5.2%. Tables are used in a similar way to those for correlation coefficient r.

For confirmation of hypothesis, χ^2 is low, the tables giving significance levels. Caution is needed if a very small value of chi-squared appears, however.

Null hypothesis assumes that no disturbing effects are operating. If in fact there are such effects then chi-squared will be large. This test cannot prove or disprove your predictions; it can only show, to significance levels which you select, that your results are either consistent or inconsistent with these predictions.

Exercise D

1. Theory predicts that in throwing three coins the frequencies of the appearance of 0, 1, 2, 3 heads will be in the ratio 1:3:3:1. In a run of 144 such throws the categories 0, 1, 2, 3 turned out to contain 17, 60, 46, 21 respectively. Is this consistent with the stated hypothesis?
2. The effectiveness of two insect-repellents was tested using 150 people; 50 applied type *A* to their skins, 50 used type *B*, while 50 acted as untreated controls. At the end of the day the results were as follows:

	Repellent A	*Repellent B*	*No repellent*
Bitten	18	27	33
Not bitten	32	23	17

   How effective do you consider each repellent to be? (Notice that you will have to perform two separate χ^2 analyses, type *A* with control, then type *B* with control. You cannot combine the two into one contingency table because the results for *A* and *B* are independent).
3. The number of parasites living on the freshwater shrimp, *Gammarus pulex*, has been investigated in colonies living in water travelling at different speeds. Is the degree of parasitism affected by the speed of the current?

	Number of animals	
	Parasitized	*Not parasitized*
Still water	125	101
Slow current	94	104
Rapid current	69	107

SAMPLING TECHNIQUES

When engaged in fieldwork you will almost always be following instructions, acting as part of a team and responsible perhaps for a small part of the whole project being undertaken. Nevertheless you may want to plan an investigation for yourself, and in any case your work will be the better for some understanding of sampling techniques, both in avoiding serious mistakes and in assessing your own and other people's results in an informed way. Specialised texts give detailed instruction; we confine ourselves to giving a few notes on sampling and random numbers.

SIMPLE SAMPLES

Are desirable theoretically; if you can use them, do, but as we said before samples without replacement can be treated in the same way as long as the sample size is small compared with the population.

SAMPLE SIZE

Is usually predetermined by such factors as the time, energy, money and material available, but good planning before you start is the only way in which you can be sure of extracting the maximum possible amount of information from your sample. It is better to have a small sample which you have planned to be as little biased as possible than a large one suspect in this respect. Suppose for instance that a research worker has to obtain viewing figures for a Rugby International televised on a Saturday afternoon, by interviewing people. If the sample consists of the first 100 people met outside a supermarket at 11.00 hours on Monday morning it is unlikely to yield useful information—except perhaps that housewives are not rugger enthusiasts. 50 people chosen to represent a cross-section of the public (not at all an easy matter) would provide better results.

Your aim is to obtain as large a sample as possible while avoiding bias; reject any method which gives a large sample if you suspect that it will be biased. We now describe some very simple methods which will help.

METHODS OF SAMPLING

The secret of success is to decide how to choose each element *before you begin*, and then to stick to your chosen procedure. This sounds simpler than it is; if you have ever peeled potatoes you will know how easy it is to pick all the large, unblemished ones first without the least intention of doing so. Suppose you need to analyse the composition of trees in a large wood where you cannot examine every one. You could decide on a line, and record the species of every third tree along it; on the same line you could choose the trees to be examined using random numbers. Either method should give good results so long as you accept what comes; of course a stream or an outcrop of rock may interrupt your steady progress.

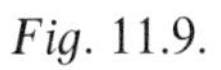
Fig. 11.9.

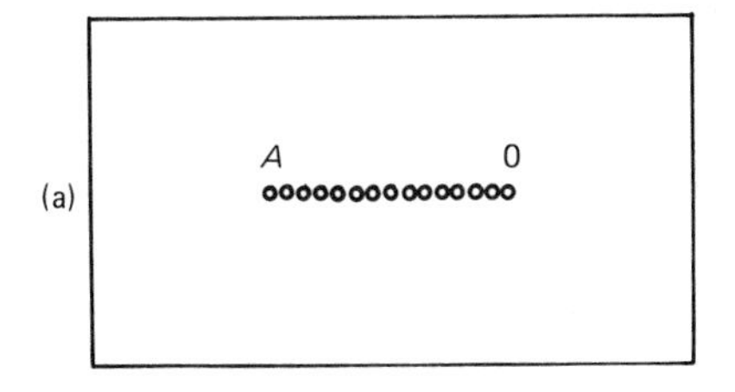

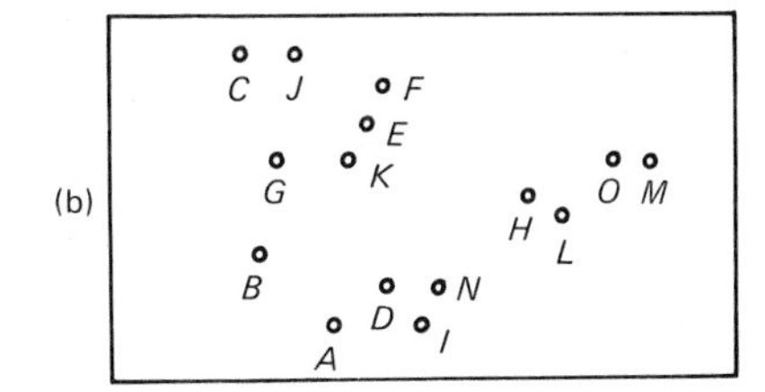

RANDOM NUMBERS

It is surprisingly difficult to generate random numbers. If you try to do it by writing down numbers, or by asking a group of people to call them out, it is most unlikely that the list will contain truly random numbers; you may favour a number without realising it, or you find that you are actually forming a distribution of digits which is too uniform. The tables of random numbers provide the simplest way out of this difficulty, and one copy of such a table, taken with you on a field course, is very helpful.

A telephone directory is a good source of random numbers. Open at random, start at the top of the page and choose the last two digits of each telephone number (sometimes there is a code in certain districts—in ours many numbers begin with 61—which makes it unwise to take the first digit).

If you have access to a computer or a programmable calculator you may find that there is a program which will print out, or in the case of the calculator, show, random numbers.

Without a table you can throw a conventional die for digits from 1 to 6 (do not use two dice to try to generate the numbers 2 to 12; the scores are not equally probable), and the time-honoured toss of a coin can always be used when there are only two alternatives.

A quick method of distributing students using quadrats involves random numbers taken in pairs. The students line up side by side across the centre of the area to be studied, figure 11.9(a); each is then given two random numbers (read off from the table, starting from any point in it). The first indicates the number of paces forward (if even) or backward (if odd). The second gives the number of paces to the right (if even) or to the left (if odd). Figure 11.9(b) shows what happened in one case. Variants will occur to you, adapted to the particular field study of the moment.

For good samples, think before you begin, decide on a method which will reduce bias as much as possible, and stick to it throughout your experiment.

APPENDIX A

S.I. UNITS

There are seven basic units in the system (Système Internationale d'Unités).

Physical quantity	*Unit*	*Symbol*
Mass	kilogram	kg
Length	metre	m
Time	second	s
Thermodynamic temperature	kelvin	K
Electric current	ampere	A
Amount of substance	mole	mol
Luminous intensity	candela	cd

SUBDIVISION AND MULTIPLES OF METRIC UNITS

Power of ten	*Prefix*	*Symbol*
10^{-12}	pico	p
10^{-9}	nano	n
10^{-6}	micro	μ
10^{-3}	milli	m
10^{-2}	centi	c
10^{-1}	deci	d
10^{3}	kilo	k
10^{6}	mega	M
10^{9}	giga	G
10^{12}	tera	T

APPENDIX B

ORDERS OF MAGNITUDE

Multiple of the metre, m	S.I. units	Resolving power of microscopes	Average dimensions of biological structures: Dimensions in standard form	Dimensions in S.I. units	
m	m metre		3.0×10^{0} m 1.7×10^{0} m	3.0 m 1.7 m	Height of elephant Height of man
$m \times 10^{-1}$	dm decimetre		6.0×10^{-1} m 1.1×10^{-1} m	600 mm 110 mm	Length of gull's wing Length of sparrow's wing
$m \times 10^{-2}$	cm centimetre		5.0×10^{-2} m 4.4×10^{-2} m	50 mm 44 mm	Length of striated muscle cell Length of 3 month old human foetus
$m \times 10^{-3}$	mm millimetre		3.5×10^{-3} m 9.0×10^{-3} m	3.5 mm 9.0 mm	Length of 1 month old human embryo Length of housefly
$m \times 10^{-4}$			1.5×10^{-4} m 1.0×10^{-4} m	150 μm 100 μm	Length of flagellum Diameter of human egg cell
$m \times 10^{-5}$			7.5×10^{-5} m 1.0×10^{-5} m	75 μm 10 μm	Diameter of striated muscle cell Length of chloroplast
$m \times 10^{-6}$	μm micrometre*		7.5×10^{-6} m 2.0×10^{-6} m	7.5 μm 2.0 μm	Length of cilium Diameter of bacterium
$m \times 10^{-7}$		Light microscope	5.0×10^{-7} m 1.5×10^{-7} m	500 nm 150 nm	Diameter of chloroplast Diameter of centriole
$m \times 10^{-8}$			1.5×10^{-8} m 1.0×10^{-8} m	15 nm 10 nm	Diameter of ribosome Diameter of small virus
$m \times 10^{-9}$	nm nanometre*	Electronmicroscope for biological material	7.0×10^{-9} m 5.0×10^{-9} m	7 nm 5 nm	Thickness of plasma membrane Diameter of haemoglobin molecule
$m \times 10^{-10}$	(Å)†	Electronmicroscope for crystals	1.54×10^{-10} m 1.50×10^{-10} m	154 pm 150 pm	Wavelength of X-rays commonly used in diffraction Distance between 2 single-bonded carbon atoms
$m \times 10^{-12}$	pm picometre				

* Older books may refer to μm simply as μ (microns) and to nm as mμ (millimicrons).
† Ångstroms are not S.I. units but will be found in older texts.

APPENDIX C

The correlation coefficient, r

Degrees of freedom	*Probability, p*				
	0.1	0.05	0.02	0.01	0.001
1	0.988	0.997	1.000	1.000	1.000
2	0.900	0.950	0.980	0.990	0.999
3	0.805	0.878	0.934	0.959	0.991
4	0.729	0.811	0.882	0.917	0.974
5	0.669	0.755	0.833	0.875	0.951
6	0.622	0.707	0.789	0.834	0.925
7	0.582	0.666	0.750	0.798	0.898
8	0.549	0.632	0.716	0.765	0.872
9	0.521	0.602	0.685	0.735	0.847
10	0.497	0.576	0.658	0.708	0.823
11	0.476	0.553	0.634	0.684	0.801
12	0.458	0.532	0.612	0.661	0.780
13	0.441	0.514	0.592	0.641	0.760
14	0.426	0.497	0.574	0.623	0.742
15	0.412	0.482	0.558	0.606	0.725
16	0.400	0.468	0.543	0.590	0.708
17	0.389	0.456	0.529	0.575	0.693
18	0.378	0.444	0.516	0.561	0.679
19	0.369	0.433	0.503	0.549	0.665
20	0.360	0.423	0.492	0.537	0.652
25	0.323	0.381	0.445	0.487	0.597
30	0.296	0.349	0.409	0.449	0.554
35	0.275	0.325	0.381	0.418	0.519
40	0.257	0.304	0.358	0.393	0.490
45	0.243	0.288	0.338	0.372	0.465
50	0.231	0.273	0.322	0.354	0.443
60	0.211	0.250	0.295	0.325	0.408
70	0.195	0.232	0.274	0.302	0.380
80	0.183	0.217	0.257	0.283	0.357
90	0.173	0.205	0.242	0.267	0.338
100	0.164	0.195	0.230	0.254	0.321

The table is abridged from Table VII of Fisher and Yates: *Statistical Tables for Biological, Agricultural and Medical Research*, published by Longman Group Ltd., London, (previously published by Oliver and Boyd, Edinburgh), and by permission of the authors and publishers.

APPENDIX D

TABLE OF RANDOM NUMBERS

05729	03242	12742	52751	55749
54212	44935	72440	44186	11385
08290	48552	01920	61676	90835
81448	26985	27525	24915	96178
10049	24459	42580	74043	00613
84318	61517	19325	09166	10516
06261	13662	58616	45837	02481
14690	25385	92501	43571	81489
71505	47218	90136	98674	72739
97946	97017	06492	01610	16265
69108	19675	17381	38176	11231
43543	18129	28818	10058	78496
74157	07953	30797	14749	35244
77674	59616	46573	04347	44678
70014	16314	09852	44394	75604

APPENDIX E

THE NORMAL PROBABILITY FUNCTION

x	0	1	2	3	4	5	6	7	8	9
0.0	.500	.504	.508	.512	.516	.520	.524	.528	.532	.536
0.1	.540	.544	.548	.552	.556	.560	.564	.567	.571	.575
0.2	.579	.583	.587	.591	.595	.599	.603	.606	.610	.614
0.3	.618	.622	.626	.629	.633	.637	.641	.644	.648	.652
0.4	.655	.659	.663	.666	.670	.674	.677	.681	.684	.688
0.5	.691	.695	.698	.702	.705	.709	.712	.716	.719	.722
0.6	.726	.729	.732	.736	.739	.742	.745	.749	.752	.755
0.7	.758	.761	.764	.767	.770	.773	.776	.779	.782	.785
0.8	.788	.791	.794	.797	.800	.802	.805	.808	.811	.813
0.9	.816	.819	.821	.824	.826	.829	.831	.834	.836	.839
1.0	.841	.844	.846	.848	.851	.853	.855	.858	.860	.862
1.1	.864	.867	.869	.871	.873	.875	.877	.879	.881	.883
1.2	.885	.887	.889	.891	.893	.894	.896	.898	.900	.901
1.3	.903	.905	.907	.908	.910	.911	.913	.915	.916	.918
1.4	.919	.921	.922	.924	.925	.926	.928	.929	.931	.932
1.5	.933	.934	.936	.937	.938	.939	.941	.942	.943	.944
1.6	.945	.946	.947	.948	.949	.951	.952	.953	.954	.954
1.7	.955	.956	.957	.958	.959	.960	.961	.962	.962	.963
1.8	.964	.965	.966	.966	.967	.968	.969	.969	.970	.971
1.9	.971	.972	.973	.973	.974	.974	.975	.976	.976	.977
2.0	.977	.978	.978	.979	.979	.980	.980	.981	.981	.982
2.1	.982	.983	.983	.983	.984	.984	.985	.985	.985	.986
2.2	.986	.986	.987	.987	.987	.988	.988	.988	.989	.989
2.3	.9893	.9896	.9898	.9901	.9904	.9906	.9909	.9911	.9913	.9916
2.4	.9918	.9920	.9922	.9925	.9927	.9929	.9931	.9932	.9934	.9936
2.5	.9938	.9940	.9941	.9943	.9945	.9946	.9948	.9949	.9951	.9952
2.6	.9953	.9955	.9956	.9957	.9959	.9960	.9961	.9962	.9963	.9964
2.7	.9965	.9966	.9967	.9968	.9969	.9970	.9971	.9972	.9973	.9974
2.8	.9974	.9975	.9976	.9977	.9977	.9978	.9979	.9979	.9980	.9981
2.9	.9981	.9982	.9982	.9983	.9984	.9984	.9985	.9985	.9986	.9986

This table was compiled, using a TI57 programmable calculator, by Ian Gent. It tabulates values of

$$\frac{1}{\sqrt{2\pi}}\int_{-\infty}^{x} e^{-\frac{1}{2}t^2}\,dt \quad \text{for} \quad 0.5 \leqslant x \leqslant 2.99$$

at intervals of 0.01.

APPENDIX F

Distribution of χ^2

Degrees of freedom	*Probability, p* 0.99	0.98	0.95	0.90	0.80	0.50	0.20	0.10	0.05	0.02	0.01	0.001
1	0.000	0.001	0.004	0.016	0.064	0.455	1.64	2.71	3.84	5.41	6.64	10.83
2	0.020	0.040	0.103	0.211	0.446	1.386	3.22	4.61	5.99	7.82	9.21	13.82
3	0.115	0.185	0.352	0.584	1.005	2.366	4.64	6.25	7.82	9.84	11.35	16.27
4	0.297	0.429	0.711	1.064	1.649	3.357	5.99	7.78	9.49	11.67	13.28	18.47
5	0.554	0.752	1.145	1.610	2.343	4.351	7.29	9.24	11.07	13.39	15.09	20.52
6	0.872	1.134	1.635	2.204	3.070	5.35	8.56	10.65	12.59	15.03	16.81	22.46
7	1.239	1.564	2.167	2.833	3.822	6.35	9.80	12.02	14.07	16.62	18.48	24.32
8	1.646	2.032	2.733	3.490	4.594	7.34	11.03	13.36	15.51	18.17	20.09	26.13
9	2.088	2.532	3.325	4.168	5.380	8.34	12.24	14.68	16.92	19.68	21.67	27.88
10	2.558	3.059	3.940	4.865	6.179	9.34	13.44	15.99	18.31	21.16	23.21	29.59
11	3.05	3.61	4.58	5.58	6.99	10.34	14.63	17.28	19.68	22.62	24.73	31.26
12	3.57	4.18	5.23	6.30	7.81	11.34	15.81	18.55	21.03	24.05	26.22	32.91
13	4.11	4.77	5.89	7.04	8.63	12.34	16.99	19.81	22.36	25.47	27.69	34.53
14	4.66	5.37	6.57	7.79	9.47	13.34	18.15	21.06	23.69	26.87	29.14	36.12
15	5.23	5.99	7.26	8.55	10.31	14.34	19.31	22.31	25.00	28.26	30.58	37.70
16	5.81	6.61	7.96	9.31	11.15	15.34	20.47	23.54	26.30	29.63	32.00	39.25
17	6.41	7.26	8.67	10.09	12.00	16.34	21.62	24.77	27.59	31.00	33.41	40.79
18	7.02	7.91	9.39	10.87	12.86	17.34	22.76	25.99	28.87	32.35	34.81	42.31
19	7.63	8.57	10.12	11.65	13.72	18.34	23.90	27.20	30.14	33.69	36.19	43.82
20	8.26	9.24	10.85	12.44	14.58	19.34	25.04	28.41	31.41	35.02	37.57	45.32
21	8.90	9.92	11.59	13.24	15.45	20.34	26.17	29.62	32.67	36.34	38.93	46.80
22	9.54	10.60	12.34	14.04	16.31	21.34	27.30	30.81	33.92	37.66	40.29	48.27
23	10.20	11.29	13.09	14.85	17.19	22.34	28.43	32.01	35.17	38.97	41.64	49.73
24	10.86	11.99	13.85	15.66	18.06	23.34	29.55	33.20	36.42	40.27	42.98	51.18
25	11.52	12.70	14.61	16.47	18.94	24.34	30.68	34.38	37.65	41.57	44.31	52.62
26	12.20	13.41	15.38	17.29	19.82	25.34	31.80	35.56	38.89	42.86	45.64	54.05
27	12.88	14.13	16.15	18.11	20.70	26.34	32.91	36.74	40.11	44.14	46.96	55.48
28	13.57	14.85	16.93	18.94	21.59	27.34	34.03	37.92	41.34	45.42	48.28	56.89
29	14.26	15.57	17.71	19.77	22.48	28.34	35.14	39.09	42.56	46.69	49.59	58.30
30	14.95	16.31	18.49	20.60	23.36	29.34	36.25	40.26	43.77	47.96	50.89	59.70

The table is abridged from Tables IV of Fisher and Yates: *Statistical Tables for Biological, Agricultural and Medical Research*, published by Longman Group Ltd., London, (previously published by Oliver and Boyd, Edinburgh), and by permission of the authors and publishers.

ANSWERS

CHAPTER 1 AIDS TO CALCULATION

Exercise A

1. 98.11; **2.** 1250; **3.** 3.13×10^{-2};
4. 1.519×10^{5}; **5.** 3.036; **6.** 1.81×10^{1};
7. 1.028; **8.** 2.255×10^{1}; **9.** 5.971×10^{-1};
10. 9.070×10^{1}; **11.** 4.541; **12.** 1.15×10^{-1};
13. 5.89; **14.** 1.17; **15.** 4.2×10^{-10};
16. 4.82×10^{-2}; **17.** 4.6×10^{-1};
18. 0.4972, 0.5899, −0.7077, −2.3261, −1.4191, −0.6021, −2.7782, −4, −0.0458, −0.3802.

Exercise B

1. 7.09×10^{-11} m.
2. (a) 5.6 dm^3; (b) 8.0 dm^3; (c) 2.10×10^{-2} dm^3.
3. 3.1×10^{-9} mol l^{-1}.
4. $x_A = \frac{1}{6}$; $x_B = \frac{5}{6}$; $P_B = 100$ mm Hg.
5. 224.
6. 1.93×10^{5} coulombs.
7. +3.
8. 974 cm^3.
9. 64.37 cm Hg.
10. 44.5 cm^3.
11. 6.3×10^{7} N.
12. 2.01 s.
13. 5.8×10^{5} ms^{-1}.
14. (a) 95; (b) 108; (c) 95; (d) 94; (e) 104.
15. (a) 16.9; (b) 15.0; (c) 13.5; (d) 12.3; (e) 11.2.
16. (a) 61.8 g; (b) 416 g.
17. 74 g.

CHAPTER 3 GROWTH, DECAY AND LOGARITHMS

Exercise A

1. 0.125; 81; 1; 0.278; 0.0625.
3. (a) 3^3; (b) 2^6; (c) 4^3; (d) 8^2; (e) 5^4.
4. (a) 3; (b) 6; (c) 3; (d) 2; (e) 4.
5. 2^x; $x = 1.0000, 1.5850, 2.7004, 3.1699, -0.3219, -1.7370$.
e^x; $x = 0.6931, 1.0986, 1.8718, 2.1972, -0.2231, -1.2040$.
10^x; $x = 0.3010, 0.4771, 0.8129, 0.9542, -0.0969, -0.5229$.
6. (a) $9 = 3^2$;
(b) $1.699 = \log_{10} 50$;
(c) $5^{-1} = 0.2$.

Exercise B

2.9957	4.6052	−0.2232	−0.6931
20.00	100.0	0.8000	0.5000

Exercise C

(a) 4.00; (b) 5.24; (c) 2.70; (d) 14.00; (e) 1.52; (f) 0.70; (g) 4.15; (h) 3.04; (i) 1.36; (j) 0.80.

Exercise D

(a) 10^{-12} mol dm^{-3}, 0.01; (b) 8×10^{-1}, 1.25×10^{-14};
(c) 2×10^{-6}, 5.0×10^{-9}; (d) 2.5×10^{-9}, 4×10^{-6};
(e) 5×10^{-3}, 2×10^{-12}; (f) 2.5×10^{-11}, 4×10^{-4};
(g) 10^{1}, 1×10^{-15}; (h) 8×10^{2}, 1.25×10^{-17};
(i) 4×10^{-5}, 2.5×10^{-10}; (j) 2×10^{-4}, 5×10^{-11}.

CHAPTER 4 GRAPHS

Exercise A

1. C = 0.8 F − 25.6,
   C = K − 273.2, intercept on K = 0 axis is
   C = −273°C, absolute zero.
   C, K and F are temperatures measured in °C, K and °F respectively.
2. Maximum density = 1000.00 kg m^{-3} at 4°C.
3. $I = 15$ V.
4. $e = 3.3$ L, range $0 \leqslant L \leqslant 100$ N.
5. The smooth increase in boiling point with increasing number of carbon atoms can be attributed to the increasing forces of attraction between molecules of increasing size.
6. The melting points decrease as the atomic number increases, which is a typical feature of metallic bonding. The outermost *s*-electrons in these metals are weakly held by the nucleus particularly as the atomic radius increases.
8. The result for 25 units probably doesn't indicate a change in gradient since the volume of bubbles is not constant so small variations in number are not significant.
9. (a) At the higher light intensity, the rate increases by a factor of 1.8. At the lower light intensity, a rise in temperature has little effect. (The early stages of photosynthesis are photochemical. If these are slow, the thermochemical reactions which follow cannot take advantage of the rise in temperature.)
   (b) Suggested intensities: 0.5×10^{-4}, 2×10^{-4}, 3×10^{-4} units.
   Distance in cm: 141, 70.7, 57.7.

Exercise B

For questions 1 to 7 the y intercept is given first followed by the gradient.

1. 5, 3; **2.** 0, 2; **3.** 4, $\frac{1}{3}$;
4. $-\frac{5}{3}, \frac{7}{3}$; **5.** 4, −1; **6.** −8, 2.
7. No intercept, 0.
8. $V = 0.39T + 1.06$.
   $V = 0$ at $T = -272$°C.

9. Gradient = 0.0022 ohm °C^{-1}.
Resistance at $T = 0$°C is 0.51 Ω.
10. $t = -1.1\,x$.

Exercise C

1. log V against log P is linear, with gradient -1; $V \propto \frac{1}{P}$.
2. Either log y against log x
or y against $1/x$, gradient $= 234 = (15.3)^2$;
$xy = f^2$.
3. Either log f against log l
or f against $1/l$, $f \propto 1/l$.
4. log R against log d; gradient $= -2$, $R \infty 1/d^2$.
5. Intercepts on $1/v$ and $1/u$ axes are 0.0655 cm^{-1},
$1/f = 0.0655$ cm$^{-1} \Rightarrow f = 15.3$ cm.
6. l against $1/f$.
Intercept on l axis $= -1$ cm;
Gradient $= 8.5 \times 10^4$ cm s^{-1};
Speed of sound in air = 34000 cm s^{-1}.
7. $x = 2$ cm.
8. The steep initial gradient is due to quantitative precipitation of $Ag^+ + Cl^- \rightarrow AgCl$. The flat portion occurs because all the Ag^+ has been precipitated out of solution. The loss of weight may be due to experimental error and/or photochemical decomposition.
9. (a) $\log_{10}k = \text{constant} - E/2.303\,RT$.
(b) Plot $\log_{10}k$ against $1/T$ and use the gradient to calculate E.
10. (a) $\Delta H_{Vap} \approx 40$ kJ mol^{-1};
(b) 343 K. The boiling point of a liquid occurs when the saturated vapour pressure equals the external pressure.

Exercise D

1. A. The reaction has gone to completion and no more $CaCl_2$ is being formed despite the addition of CaO.
2. (a) A; (b) B; (c) A; (d) approximately 50 g per 100 g water.
3. (a) Triple point -90°C, 0.9 atm. Phase boundaries.
(b) For temperatures greater than -90°C.
(c) Approximately 1.1.
4. (a) Z.
(b) The reaction has gone to completion.
(c) 1 minute.
(d) The rate would be faster as the surface area of the solid reactant has been increased.
5. The shape of this orbital is a sphere with the maximum probability of finding the electron at the centre. The boundary surface is not strictly defined.
6. (a) (i) 7 16 25 35
(ii) 7 16 30 46
(c) 28.

7. The most stable elements with maximum binding energies lie between chromium and zinc.
8. (iv) $8.31\,\mathrm{J\,mol^{-1}\,K^{-1}}$.
9. A scale of 300 to $400\,\mathrm{cm^3}$ is sufficient for the volume axis
(b) $311\,\mathrm{cm^3}$; (c) 31.9; (d) $1.14\,\mathrm{cm^3K^{-1}}$; (e) 273 K;
(f) Gases expand or contract by $\frac{1}{273}$ of their volume at 0°C for every °C that the temperature is raised or lowered respectively—Charles' law.

CHAPTER 5 ALGEBRAIC TECHNIQUES

Exercise A

1. $n = (\sigma/S)^2$; **2.** $u = v/m$; **3.** $c = df_m$;
4. (i) $\varepsilon_0 = 4\pi r^2 F/qQ$; (ii) $r = \sqrt{qQ/4\pi\varepsilon_0 F}$.
5. $R = R_1R_2/(R_1 + R_2)$.
6. $c = 8\pi lma^2/(a + b)\,y$.
7. (i) $T_1 = P_1V_1T_2/P_2V_2$; (ii) $V_1 = P_2V_2T_1/P_1T_2$.
8. $\lambda = n^2m^2/R_H(m^2 - n^2)$.
9. 0.30 m; **10.** $1.29\,\mathrm{cm^2}$.

Exercise B

1. 1, $-\frac{1}{2}$; **2.** 2, -2; **3.** 1, 2;
4. No solution; **5.** 2, 3; **6.** -1, 3;
7. 13/14, $-11/14$; **8.** 6, 1; **9.** 11, 4.
10. $\lambda = 5.91 \times 10^{-5}$ cm, $n = 6$.
11. $\lambda = 5.74 \times 10^{-5}$ cm, $n = 4$.
12. $\lambda = 4.35 \times 10^{-5}$ cm, $n = 3$.

Exercise C

1. 0, $\frac{2}{3}$; **2.** 1, 2; **3.** 5,2;
4. 5, -1; **5.** -5, 2; **6.** $\frac{1}{2}$, 4;
7. $\frac{2}{3}$, -5; **8.** 4, 2.

Exercise D

1. 3.62, 1.38; **2.** -4, 7; **3.** $\frac{5}{2}$, -3;
4. No real roots; **5.** -2, -1; **6.** No real roots;
7. 0.73, -0.23; **8.** $\frac{1}{3}$, -1; **9.** 2.62, 0.38;
10. 0.46, -0.86; **11.** $-1\frac{1}{2}$; **12.** 1.19, -4.19;
13. 0, $3\frac{1}{3}$; **14.** 5, -6.

Exercise E

1. (i) $2(a-x) \quad x \quad 3x$
$K = 27x^4/4(a-x)^2V^2$ where V is the volume at equilibrium.
(ii) $(a-x) \quad (b-x) \quad x \quad x$
$K = x^2/(a-x)(b-x)$.
(iii) 2 moles.
2. 2.232×10^{-2} moles.
3. 45.7%.

CHAPTER 6 STATISTICS

Exercise A

1. 1.629; **2.** 1.620, 1.55 – 1.59.

Exercise B

1. 2.01, 2; **2.** 62.6;
3. 4.98, 4; **4.** 43, 53, 28, 23.

Exercise C

1. 25, 9, 360, 61.
2. (a) 14, (b) 20.
3. 12.
4. (a) $\frac{1}{N}\sum_{i=1}^{n} f_i x_i$, (b) $\frac{1}{N}\sum_{i=1}^{N} x_i$.

Exercise D

1. 1.79, 1.34; **2.** 68.03, 8.25; **3.** 1.66, 1.29;
4. 278.2, 16.7; **6.** 2, 1.80, 1.34; **7.** 2.37, 1.83.

CHAPTER 7 TRIGONOMETRY

Exercise A

1. 0.75, 0.37, 1.22, 3.45, 4.24, 5.64.
2. 57.30, 28.65, 38.39, 80.22, 148.98, 194.82.
3. $\pi/6$, $\pi/3$, $2\pi/3$, $5\pi/6$, $7\pi/6$, $4\pi/3$, $5\pi/3$, $11\pi/6$, $\pi/4$, $3\pi/4$, $5\pi/4$, $7\pi/4$, $\pi/2$, π, $3\pi/2$, 2π.
4. 22.5°, 56.3°, 27.0°, 157.5°, 17.2°, 114.6°, 322.0°.

Exercise B

1. 1.54, 1.52, 1.53, mean value = 1.53.
2. 24.4°, 41.1°, 37.3°, 43.2°.

Exercise C

1. (a) 38.45 rad or 2203°, (b) 0.0384 rad or 2.203°.
2. 0.3536, 0.5000, 0.3536, 0.0000, −0.5000, 0.0000.
3. (a) 10.0 N vertically, 17.3 N horizontally.
(b) 0.0129 N vertically, 0.0483 N horizontally.
(c) 220 N vertically, 0 N horizontally.

CHAPTER 8 CALCULUS; DIFFERENTIATION

Exercise A

1. Rate of cooling is directly proportional to excess temperature.
2. 1.21, 6.25, 4.004, 9, 10.24.
3. 2.1, 4.5, 6.2.
4. 3.65; gradient from graph = 3.0.
5. $3x^2$; 27, 0, 3.
6. (b) 20% hydrolysed per minute at 15°C;
24% at 25°C; 42% at 35°C;
46% at 45°C;
(c) 15°C to 25°C, 1.20; 25°C to 35°C, 1.75;
35°C to 45°C, 1.10;
(d) 25°C to 35°C.

Exercise B

1. $5x^4, -3x^{-4}, \frac{1}{2}x^{-1/2}$.
2. $dy/dx = 3x^2$; 12, 0, 12, 75.
3. $dy/dx = x^{-2} = 1/x^2$; 4, $\frac{1}{9}$, 4, $\frac{1}{9}$.
4. 2, 0.
5. $2\cos x - 3\sin x$.
6. 0, −0.8660, −1.0000, −0.7071.
7. $E_0 = BA\omega$.
8. $x = 8$.
9. $\omega t_{min} = \pi$, $t_{min} = \pi/\omega$.
10. $y = -71$, minimum; $y = 48$, maximum.

CHAPTER 9 CALCULUS; INTEGRATION

Exercise A

1. (i) 88, (ii) 12, (iii) −9.17.
2. 202.0 m
3. (a) 1.7899, (b) 0.0000. The area under the curves $y = \sin\theta$ and $y = \cos\theta$ for $0 \leqslant \theta \leqslant \pi/2$ is 0.8949.
4. 7.404.
5. (a) 693.7 energy units, (b) 798.8 energy units. Energy difference = 105.1 units.

Exercise B

1. (i) 12, (ii) $30\frac{1}{3}$, (iii) −0.880, (iv) π,
(v) 0.2929, (vi) 0.5108.
2. (i) 30, (ii) $20\frac{5}{6}$, (iii) 0.2929, (iv) 0, (v) 1.5,
(vi) 8/3, (vii) 6.3891, (viii) 0.0664, (ix) 1.386.
3. (i) 12, (ii) 65/3, (iii) 0.444,
(iv) $(\sqrt{3} - 1)/2 = 0.3660$, (v) $1/2[\pi/4 - 1/2]$.
5. c is the constant of integration
(i) $x^4/4 - 3x^2/2 + c$
(*ii*) $3\sin x + x + c$
(iii) $\dfrac{\sin(\omega t)}{\omega} + c$.
6. $202\frac{2}{3}$ m.

CHAPTER 10 EXPONENTIAL VARIATION

Exercise A

1. $T_2 = 3.25$ days; $T_e = 4.6$ days; $N = 100\,e^{t/4.6}$.
2. (a) 2.05, (b) 1.82.
3. (a) 6 hours to 9 hours; (b) bell-shaped; (c) growth is exponential at first (i.e. hourly increments represent a constant ratio) until factors such as limited food supply begin to have an effect.

Exercise B

1. $5.93 \times 10^{-3}\,s^{-1}$, 1.17×10^2 s.
2. 1400 years.

3. $3.26 \times 10^{-3}\,s^{-1}$.
4. $\log_{10}k$ against $1/T$.

Exercise C
1. $0.01\,s^{-1}$, 69.3 s.
2. (b) $k = 0.1002$ year^{-1}, $T_2 = 6.919$ year.
3. 1.236 cm.
4. The graph of log (count rate) against absorber thickness is linear for absorber thicknesses up to 0.06 mm.

CHAPTER 11 STATISTICS

Exercise A
1. (a) -3.33, 0.67, 0.73;
(b) 0.80, -1.60, $+2.20$;
(c) 0.89, 2.68, -2.68.
2. 8.16 to 23.84.
3. 3.4%.

Exercise B
1. (a) 4, confidence limit $<95\%$;
(b) 19, confidence limit 98%;
(c) 10, confidence limit near 90%;
(d) 29, confidence limit 95%;
(e) 89 (use 90), confidence limit close to 98%.
2. 0.7523, Yes.
3. (a) 0.8281, Yes; (b) 0.0302, No; (c) -0.8919, Yes; (d) 0.6815, Yes.

Exercise C
1. 2.799, significant; **2.** 3.205, significant.

Exercise D
1. 2.407, not significant.
2. A, 9.004, significant; B, 1.500, not significant.
3. 10.32, significant.